543 Most Important European Scientists All of Time (0000 – 2015)

Ressources of photos :
free encyclopedia wikipedia.org

Do you know the most important European scientist?
I am sure you have often heard a few names from them, But do you know how they looked?
Do you know which country these scientists came from? When were they born and died?

I must say honestly, I have also not been able to answer many of these questions.
But this book gives you answers.

Have fun!

Kennen Sie die wichtigsten europeaischen Wissenschaftler?
Ich bin überzeugt, sie haben oft manchen Namen von ihnen gehört, aber wissen Sie wie sie ausgeschaut haben? Wissen Sie, aus welchem Land diese Wissenschaftler kamen? Wann sie geboren und gestorben sind?

Ich muss ehrlich sagen, ich habe auch viele von diesen Fragen nicht beantworten können.
Aber dieses Buch gibt Ihnen Antworten

Viel Spaß!

Connaissez-vous les plus grands scientifiques européens?
Je suis sûr que vous avez souvent entendu quelques noms d'eux, Mais connaissez-vous leurs visage?
Savez-vous de quel pays étaient ces scientifique? Quand sont-ils nés et sont-ils morts?

Je dois dire honnêtement que je n'ai pas non plus réussi à répondre à bon nombre de ces questions.
Mais ce livre vous donne des réponses.

Amusez bien!

Best European Scientist

scientists	profession	country
Abel/ Niels Henrik (1802 – 1829)	mathematician	NORWAY
Adrian/ Edgar Douglas (1889 – 1977)	electrophysiologist	ENGLAND
Agassiz/ Louis (1807 – 1873)	biologist and geologist	SWITZERLAND
Agnesi/ Maria Gaetana (1718 – 1799)	mathematician and philosopher	ITALY
Ahlfors/ Lars Valerian (1907 – 1996)	mathematician, remembered for his work in the field of Riemann surfaces and his text on complex analysis	FINLAND
Ajzenberg Selove/ Fay (1926 – 2012)	nuclear physicist	GERMANY
Aleksandrov/ Aleksandr Danilovich (1912 – 1999)	mathematician, physicist, philosopher and mountaineer.	RUSSIA
Alexandria/ Diophantus (AD 201 – AD 285)	mathematician	GREECE
Alfvén/ Hannes Olof Gösta (1908 – 1995)	electrical engineer, plasma physicist	SWEDEN
Allais/ Maurice Félix Charles (1911 – 2010)	economist	FRANCE
Ampère/ Andre Marie (1775 – 1836)	physicist and mathematician	FRANCE
Anning/ Mary (1799 – 1847)	fossil collector, dealer, and palaeontologist	ENGLAND
Appellöf/ Jakob Johan Adolf (1857 – 1921)	marine zoologist	SWEDEN
Appleton/ Edward Victor (1892 – 1965)	physicist	ENGLAND
Arber/ Agnes (1879 – 1960)	plant morphologist and anatomist, historian of botanyand philosopher of biology	ENGLAND
Arrhenius/ Svante (1859 – 1927)	scientist, physicist, chemist	SWEDEN

Best European Scientist

scientists	profession	country
Asclepiades (c. 124 or 129 – 40 BC)	he was a Greek physician born at Prusa in Bithynia in Asia Minorand flourished at Rome, where he established Greek medicine near the end of the 2nd century BC	Greece
Aston/ Francis William (1877 – 1945)	chemist and physicist	ENGLAND
Avogadro/ Amedeo (1776 – 1856)	scientist	ITALY
Ayrton Hertha Marks (1854 – 1923)	engineer, mathematician, physicist, and inventor	ENGLAND
Babbage/ Charles (1791 – 1871)	mathematician, philosopher, inventor and mechanical engineer	ENGLAND
Bachet/ Claude (1581 – 1638)	mathematician, linguist	FRANCE
Bacon/ Francis (1561 – 1626)	philosopher, scientist, jurist	ENGLAND
Bacon/ Roger (1214 – 1294)	philosopher and Franciscan friar who placed considerable emphasis on the study of nature through empirical methods	ENGLAND
Baekeland/ Leo (1863 – 1944)	chemist. He invented Velox photographic paper in 1893 and Bakelite in 1907	BELGIUM
Baeyer/ Johann Friedrich Wilhelm Adolf (1835 – 1917)	chemist	GERMANY
Bain/ Alexander (1818 – 1903)	philosopher and educationalist	SCOTLAND
Banks/ Joseph (1743 – 1820)	naturalist, botanist	ENGLAND
Bárány/ Róbert (1876 – 1936)	otologist	AUSTRIA

Best European Scientist

scientists	profession	country
Barrow/ Isaac (1630 – 1677)	Christian theologian, and mathematician who is generally given credit for his early role in the development of infinitesimal calculus	ENGLAND
Bassi/ Laura (1711 – 1778)	scientist, physicist	ITALY
Bayliss/ William (1860 – 1924)	physiologist	ENGLAND
Becquerel/ Henri (1852 – 1908)	physicist	FRANCE
Behring/ Emil Adolf (1854 – 1917)	physiologist	GERMANY
Békésy/ György (1899 – 1972)	biophysicist	HUNGARY
Bell/ John (1763 – 1820)	anatomist and surgeon	SCOTLAND
Bell/ John Stewart (1928 – 1990)	physicist, and the originator of Bell's theorem, a significant theorem in quantum physics regarding hidden variable theories	IRELAND
Beltrami/ Eugenio (1835 – 1899)	mathematician	ITALY
Bentz/ Melitta (1873 – 1950)	inventor (the coffee filter), entrepreneur	GERMANY
Benz/ Carl (1844 – 1929)	engine designer and car engineer, inventor	GERMANY
Bergius/ Friedrich Karl Rudolf (1884 – 1949)	chemist	GERMANY
Berliner/ Emile (1851 – 1929)	inventor "disc record gramophone"	GERMANY
Bernard/ Claude (1813 – 1878)	physiologist	FRANCE
Bernoulli/ Daniel (1700 – 1782)	mathematician and physicist	SWITZERLAND
Bernoulli/ Jacob (1654 – 1705)	mathematician	SWITZERLAND

Best European Scientist

scientists	profession	country
Bernoulli/ Johann (1667 – 1748)	mathematician	SWITZERLAND
Bertrand/ Joseph (1822 – 1900) Contents	mathematician who worked in the fields of number theory, differential geometry, probability theory, economics and thermodynamics	FRANCE
Berzelius/ Jöns Jacob (1779 – 1848)	chemist, one of the founders of modern chemistry	SWEDEN
Besicovitch/ Abram (1891 – 1970)	mathematician	RUSSIA
Bessel/ Wilhelm (1784 – 1846)	astronomer, mathematician (systematizer of the Bessel functions), He was the first astronomer to determine the distance from the sun to another star by the method of parallax	GERMANY
Bessemer/ Henry (1813 – 1898)	engineer, inventor (process for the manufacture of steel)	ENGLAND
Bethe/ Hans (1906 – 2005)	nuclear physicist	GERMANY
Binet/ Alfred (1857 – 1911)	psychologist	FRANCE
Birkeland/ Kristian (1867 – 1917)	scientist	NORWAY
Bíro/ Laszlo Jozsef (1899 – 1985)	inventor, the ballpoint pen, still commonly called biro after him.	HUNGARY
Bjerknes/ Carl (1825 – 1903)	mathematician and physicist	NORWAY
Blackett/ Patrick Maynard Stuart (1897 – 1974)	experimental physicist	ENGLAND
Blackwell/ Elizabeth (1821 – 1910)	medicine	ENGLAND

Best European Scientist

scientists	profession	country
Bloch/ Felix (1905 – 1983)	physicist	SWITZERLAND
Bloch/ Konrad Emil (1912 – 2000)	biochemist	GERMANY
Boas/ Franz (1858 – 1942)	anthropologist	GERMANY
Bohr/ Niels (1885 – 1962)	physicist	DENMARK
Boltzmann/ Ludwig (1844 – 1906)	physicist and philosopher	AUSTRIA
Bonaventura de Cavalieri/ Francesco (1598 – 1647)	mathematician	ITALY
Boole/ George (1815 – 1864)	mathematician, philosopher and logician	ENGLAND
Bordet/ Jules Jean Baptiste Vincent (1870 – 1961)	immunologist and microbiologist	BELGIUM
Borel/ Félix Édouard Justin Émile (1871 – 1956)	mathematician	FRANCE
Born/ Max (1882 – 1970)	physicist and mathematician	GERMANY
Bosch/ Carl (1874 – 1940)	chemist and engineer	GERMANY
Bosch/ Robert (1861 – 1942)	engineer and inventor	GERMANY
Bovet/ Daniel (1907 – 1992)	pharmacologist	SWITZERLAND
Boyle/ Robert (1627 – 1691)	natural philosopher, chemist, physicist, and inventor	IRELAND
Bragg William Henry (1862 – 1942)	physicist, chemist, mathematician	ENGLAND
Bragg/ William Lawrence (1890 – 1971)	physicist and X-ray crystallographer	ENGLAND
Brahe/ Tycho (1546 – 1601)	astronomer and alchemist and has been described more recently as "the first competent mind in modern astronomy to feel ardently the passion for	DENMARK

scientists	profession	country
	exact empirical facts	
Braille/ Louis (1809 – 1852)	educator and inventor of a system of reading and writing for use by the blind or visually impaired. His system remains known worldwide simply as braille	FRANCE
Braun/ Karl Ferdinand (1850 – 1918)	inventor, physicist	GERMANY
Brongniart/ Alexandre (1770 – 1847)	chemist, mineralogist, and zoologist	FRANCE
Brønsted/ Johannes Nicolaus (1879 – 1947)	physical chemist	DENMARK
Brown/ Robert (1773 – 1858)	botanist and palaeobotanist	SCOTLAND
Brunfels/ Otto (1488 – 1534)	theologian and botanist, was the first to produce a major work on plants	GERMANY
Bruno/ Giordano (1548 – 1600)	philosopher, mathematician, poet, and astrologer. He is celebrated for his cosmological theories	ITALY
Buchner/ Eduard (1860 – 1917)	chemist and zymologist	GERMANY
Buckland/ William (1784 – 1856)	theologian, geologist and palaeontologist	ENGLAND
Bunsen/ Robert (1811 – 1899)	chemist. He investigated emission spectra of heated elements	GERMANY
Burnet/ Thomas (1635? – 1715)	theologian and writer on cosmogony	ENGLAND
Butenandt/ Adolf Friedrich Johann (1903 – 1995)	biochemist	GERMANY
Cajal/ Santiago Ramón y (1852 – 1934)	pathologist, histologist, neuroscientist	SPAIN
Cannizzaro/ Stanislao (1826 – 1910)	chemist. He is remembered today	ITALY

Best European Scientist

scientists	profession	country
	largely for the Cannizzaro reaction and for his influential role in the atomic-weight deliberations of the Karlsruhe Congress in 1860	
Cantor/ Georg (1845 – 1918)	mathematician	GERMANY
Cardano/ Girolamo (1501 – 1576)	mathematician, physician, astrologer, philosopher	ITALY
Carrel/ Alexis (1873 – 1944)	surgeon and biologist	FRANCE
Cartan/ Élie (1869 – 1951)	mathematician	FRANCE
Cartan/ Henri Paul (1904 – 2008)	mathematician	FRANCE
Cartwright/ Mary (1900 – 1998)	mathematician, she was the first to analyze a dynamical system with chaos	ENGLAND
Cauchy/ Augustin (1789 – 1857)	mathematician	FRANCE
Cayley/ Arthur (1821-1895)	mathematician	ENGLAND
Celsius/ Anders (1701 – 1744)	astronomer, physicist and mathematician	SWEDEN
Celsus/ Aulus Cornelius (c. 25 BC – c. 50 AD)	he was a Roman encyclopaedist, known for his extant medical work, De Medicina, which is believed to be the only surviving section of a much larger encyclopedia.	Greece
Chadwick/ James (1891 – 1974)	physicist	ENGLAND
Chain/ Ernst Boris (1906 – 1979)	biochemist	GERMANY
Charles/ Jacques (1746 – 1823)	inventor, scientist, mathematician, and	FRANCE

Best European Scientist

scientists	profession	country
	balloonist	
Charpak/ Georges (1924 – 2010)	physicist	POLAND
Chrysippus (c. 279 BC – c. 206 BC)	philosopher	GREECE
Chun/ Carl (1852 – 1914)	marine biologist. He graduated in zoology from the University of Leipzig	GERMANY
Clairaut/ Alexis Claude (1713 – 1765)	mathematician, astronomer, geophysicist, and intellectual	FRANCE
Clavering/ Hardy Alister (1896 – 1985)	marine biologist who was an expert on marine ecosystems and zooplankton	ENGLAND
Cockcroft/ John Douglas (1897 – 1967)	physicist	ENGLAND
Cole/ Henry (1808 – 1882)	inventor, Cole is credited with devising the concept of sending greetings cards at Christmas time, introducing the world's first commercial Christmas card in 1843	ENGLAND
Copernicus Nicolaus (1472 – 1543)	mathematician and astronomer	POLAND
Cori/ Carl Ferdinand (1896 – 1984)	biochemist and pharmacologist	CZECH REPUBLIC
Cori/ Gerty Theresa (1896 – 1957)	biochemist	CZECH REPUBLIC
Cousteau/ Jacques Yves (1910 – 1997)	researcher and an ecologist who studied the lives of underwater animals and plants	FRANCE
Cowdery/ Kendrew John (1917 – 1997)	biochemist and crystallographer	ENGLAND
Cremona/ Antonio Luigi (1830 – 1903)	mathematician	ITALY

Best European Scientist

scientists	profession	country
Crick/ Francis (1916 – 2004)	molecular biologist, biophysicist, and neuroscientist	ENGLAND
Cristofori/ Bartolomeo (1655 – 1731)	maker of musical instruments, generally regarded as the inventor of the piano	ITALY
Culpeper/ Nicholas (1616 – 1654)	botanist, herbalist, physician, and astrologer	ENGLAND
Curie/ Marie (1867 – 1934)	physicist and chemist	POLAND
Curie/ Pierre (1859 – 1906)	physicist, a pioneer in crystallography, magnetism, piezoelectricity and radioactivity	FRANCE
Curie/ Joliot Irene (1897 – 1956)	chemist , scientist	FRANCE
Curie/ Joliot Jean Frédéric (1900 – 1958)	physicist	FRANCE
Czerny/ Adalbert (1863 – 1941)	pediatrician	AUSTRIA
Da Vinci/ Leonardo (1452 – 1519)	architect, mathematician, engineer, inventor, anatomist, geologist, cartographer, botanist, writer, painter, sculptor, musician	ITALY
Daimler/ Gottlieb (1834 – 1900)	engineer, industrial designer	GERMANY
Dalén/ Nils Gustaf (1869 – 1937)	scientist, inventor	SWEDEN
Dalton/ John (1766 – 1844)	chemist, meteorologist and physicist	ENGLAND
Dam/ Henrik (1895 – 1976)	biochemist and physiologist	DENMARK
Darboux/ Jean Gaston (1842 – 1917)	mathematician	FRANCE
Darwin/ Charles (1809 – 1882)	naturalist and geologist	ENGLAND

Best European Scientist

scientists	profession	country
Dausset/ Jean Baptiste Gabriel Joachim (1916 – 2009)	immunologist	FRANCE
Davidovich Landau/ Lev (1908 – 1968)	physicist	RUSSIA
Davy/ Humphry (1778 – 1829)	chemist and inventor	ENGLAND
De Abreu Freire Egas Moniz/ António Caetano (1874 – 1955)	neurologist and the developer of cerebral angiography	PORTUGAL
De Broglie/ Louis (1892 – 1987)	physicist	FRANCE
De Coulomb/ Charles Augustin (1736 – 1806)	physicist	FRANCE
De Fermat/ Pierre (1601 – 1665)	mathematician (infinitesimal calculus, including his technique of adequality)	FRANCE
De Hevesy/ George Charles (1885 – 1966)	radiochemist	HUNGARY
Debreu/ Gérard (1921 – 2004)	economist and mathematician	FRANCE
Debye/ Peter (1884 – 1966)	physicist and physical chemist,	NETHERLANDS
Dedekind/ Julius Wilhelm Richard (1831 – 1916)	mathematician	GERMANY
Democritus (c. 460 – c. 370 BC)	influential pre-Socratic philosopher	GREECE
Déscartes/ René (1596 – 1650)	philosopher, mathematician and writer	FRANCE
Detlof Bergström/ Karl Sune (1916 – 2004)	biochemist	SWEDEN
Diels/ Otto Paul Hermann (1876 – 1954)	chemist	GERMANY
Diesel/ Rudolf Christian Karl (1858 – 1913)	inventor and mechanical engineer	GERMANY
Dirac/ Paul (1902 – 1984)	theoretical physicist	ENGLAND
Divis/ Prokop (1698 – 1765)	theologian and natural	CZECH

Best European Scientist

scientists	profession	country
	scientist, who invented the first grounded lightning rod	REPUBLIC
Dobereiner/ Johann Wolfgang (1780 – 1849)	chemist who is best known for work that foreshadowed the periodic law for the chemical elements	GERMANY
Dohrn/ Anton (1840 – 1909)	marine biologist, He had mastered not only medicine but also zoology	GERMANY
Domagk/ Gerhard Johannes Paul (1895 – 1964)	pathologist and bacteriologist	GERMANY
Du Châtelet/ Émilie (1706 – 1749)	mathematician, physicist	FRANCE
Dulbecco/ Renato (1914 – 2012)	virologist	ITALY
Eddington/ Arthur (1882 – 1944)	astronomer, physicist, and mathematician	ENGLAND
Ehrlich/ Paul (1854 – 1915)	physician and scientist who worked in the fields of hematology, immunology, and chemotherapy	GERMANY
Eijkman/ Christiaan (1858 – 1930)	physician and professor of physiology	NETHERLANDS
Einstein/ Albert (1879 – 1955)	theoretical physicist and philosopher of science	GERMANY
Einthoven/ Willem (1860 – 1927)	doctor and physiologist	NETHERLANDS
Eisenstein/ Ferdinand Gotthold Max (1823 – 1852)	mathematician	GERMANY
Eratosthenes (c. 276 BC – c. 195/194 BC)	mathematician, geographer, poet, astronomer, and music theorist	GREECE
Erdös/ Paul (1913 – 1996)	mathematician. He was	HUNGARY

Best European Scientist

scientists	profession	country
	the most prolific mathematician of the 20th century, but also known for his social practice of mathematics	
Ernst/ Pauli Wolfgang (1900 – 1958)	theoretical physicist and one of the pioneers of quantum physics	AUSTRIA
Essers/ Ilse (1898 – 1994)	engineer	GERMANY
Euler/ Leonhard (1707 – 1783)	mathematician and physicist	SWITZERLAND
Faraday/ Michael (1791 – 1867)	scientist (electromagnetism and electrochemistry)	ENGLAND
Fermi/ Enrico (1901 – 1954)	physicist	ITALY
Fielding Huxley/ Andrew (1917 – 2012)	physiologist and biophysicist	ENGLAND
Fischer/ Emil (1852 – 1919)	chemist	GERMANY
Fischer/ Hans (1881 – 1945)	organic chemist	GERMANY
Fleming/ Alexander (1881 – 1955)	biologist, pharmacologist and botanist	SCOTLAND
Forßmann/ Werner Theodor Otto (1904 – 1979)	physician	GERMANY
Foucault/ Leon (1819 – 1868)	physicist	FRANCE
Fourier/ Jean Baptiste Joseph (1768 – 1830)	mathematician and physicist	FRANCE
Franck/ James (1882 – 1964)	physicist	GERMANY
Frank/ Ilya Mikhailovich (1908 – 1990)	physicist	RUSSIA
Franklin/ Rosalind (1920 – 1958)	chemist and X-ray crystallographer	ENGLAND
Fréchet/ René Maurice (1878 – 1973)	mathematician	FRANCE
Freud/ Sigmund (1856 – 1939)	neurologist	AUSTRIA
Friese Greene/ William	portrait photographer and	ENGLAND

Best European Scientist

scientists	profession	country
(1855 – 1921)	prolific inventor. He is principally known as a pioneer in the field of motion pictures and is credited by some as the "inventor" of cinematography,	
Frobenius/ Ferdinand Georg (1849 – 1917)	mathematician	GERMANY
Fuchs/ Leonard (1501 – 1566)	physician and botanist	GERMANY
Gabor/ Dennis (1900 – 1979)	electrical engineer and physicist	HUNGARY
Galenus/ Aelius (AD 129 – c.200/c.216)	surgeon and philosopher, physician, his work was the centerpiece of traditional biology and anatomy that had lasted through the Middle Ages	GREECE
Galilei/ Galileo (1564 – 1642)	physicist, mathematician, engineer, astronomer, and philosopher	ITALY
Galois/ Évariste (1811 – 1832)	mathematician	FRANCE
Galton/ Francis (1822 – 1911)	psychologist, anthropologist, eugenicist, tropical explorer, geographer, inventor, meteorologist, proto-geneticist, psychometrician, and statistician	ENGLAND
Galvani/ Luigi (1737 – 1798)	physicist and philosopher	ITALY
Gamow/ George (1904 – 1968)	theoretical physicist and cosmologist	RUSSIA
Gauss Carl Friedrich (1777 – 1855)	mathematician	GERMANY
Gay Lussac/ Joseph Louis (1778 – 1850)	chemist and physicist. He is known mostly for two laws related to gases,	FRANCE

Best European Scientist

scientists	profession	country
	and for his work on alcohol-water mixtures, which led to the degrees Gay-Lussac used to measure alcoholic beverages in many countries	
Gelfand/ Israel Moiseevich (1913 – 2009)	mathematician	RUSSIA
Gilles de Gennes/ Pierre (1932 – 2007)	physicist	FRANCE
Glover Barkla/ Charles (1877 – 1944)	physicist	ENGLAND
Göbel/ Heinrich (1818 – 1893)	precision mechanic and inventor (Incandescent light bulb)	GERMANY
Gödel/ Kurt (1906 – 1978)	logician, mathematician, and philosopher	AUSTRIA
Goeppert Mayer/ Maria (1906 – 1972)	theoretical physicist	GERMANY
Golgi/ Camillo (1843 – 1926)	physician, pathologist, scientist	ITALY
Gottlob Frege/ Friedrich Ludwig (1848 – 1925)	mathematician, logician and philosopher	GERMANY
Gowland Hopkins/ Frederick (1861 – 1947)	biochemist	ENGLAND
Graham/ Thomas (1805 – 1869)	chemist. He developed a technique to separate crystalloids from colloids, which is called "dialysis"	SCOTLAND
Graham/ Bell Alexander (1847 – 1922)	scientist, inventor, engineer and innovator	SCOTLAND
Granger/ Clive William John (1934 – 2009)	economist	WALES
Granit/ Ragnar Arthur (1900 – 1991)	scientist (Physiology, Medicine)	FINLAND

Best European Scientist

scientists	profession	country
Grassmann/ Hermann Günter (1809 – 1877)	mathematician, physicist, neohumanist, general scholar, publisher	GERMANY
Gregory/ James (1638 – 1675)	mathematician and astronomer	SCOTLAND
Grib Fibiger/ Johannes Andreas (1867 – 1928)	scientist, physician, and professor of pathological anatomy	DENMARK
Grignard/ François Auguste Victor (1871 – 1935)	chemist	FRANCE
Guillaume/ Charles Édouard (1861 – 1938)	physicist	SWITZERLAND
Gullstrand/ Allvar (1862 – 1930)	ophthalmologist and optician	SWEDEN
Gutenberg/ Johannes (1398 – 1468)	blacksmith, goldsmith, printer, and publisher	GERMANY
Györgyi/ Albert Szent (1893 – 1986)	physiologist	HUNGARY
Haber/ Fritz (1868 – 1934)	chemist	GERMANY
Hadamard/ Jacques Salomon (1865 – 1963)	mathematician	FRANCE
Haeckel/ Ernst (1834 – 1919)	biologist, naturalist, philosopher, physician	GERMANY
Hahn/ Otto (1879 – 1968)	chemist and pioneer in the fields of radioactivity and radiochemistry	GERMANY
Hahnemann/ Samuel (1755 – 1843)	physician, best known for creating a system of alternative medicine called homeopathy	GERMANY
Hallett Dale/ Henry (1875 – 1968)	pharmacologist and physiologist	ENGLAND
Halley/ Edmund (1656 – 1742)	astronomer, geophysicist, mathematician, meteorologist, and physicist	ENGLAND
Hamilton/ William Rowan (1805 – 1865)	physicist, astronomer, and mathematician, who made important	IRELAND

Best European Scientist

scientists	profession	country
	contributions to classical mechanics, optics, and algebra	
Harden/ Arthur (1865 – 1940)	biochemist	ENGLAND
Hardy/ Godfrey Harold (1877 – 1947)	mathematician	ENGLAND
Harsanyi/ John Charles (1920 – 2000)	economist	HUNGARY
Harvey/ William (1578 – 1657)	physician	ENGLAND
Hassel/ Odd (1897 – 1981)	physical chemist	NORWAY
Hausdorff/ Felix (1868 – 1942)	mathematician	GERMANY
Haxel/ Otto (1909 – 1998)	nuclear physicist	GERMANY
Hayek/ Friedrich (1899 – 1992)	economist	AUSTRIA
Heaviside/ Oliver (1850 – 1925)	electrical engineer, mathematician, and physicist	ENGLAND
Heisenberg/ Werner (1901 – 1976)	theoretical physicist and one of the key creators of quantum mechanics	GERMANY
Henlein/ Peter (1485 – 1542)	locksmith and clockmaker of Nuremberg, Germany, is often considered the inventor of the watch	GERMANY
Hermite/ Charles (1822 – 1901)	mathematician	FRANCE
Herschel/ William (1738 – 1822)	astronomer	GERMANY
Hertz/ Gustav Ludwig (1887 – 1975)	experimental physicist	GERMANY
Hertz/ Heinrich (1857 – 1894)	physicist	GERMANY
Hess/ Germain (1802 – 1850)	chemist and doctor who formulated Hess's law, an early principle of	SWITZERLAND

Best European Scientist

scientists	profession	country
	thermochemistry	
Hess/ Victor Francis (1883 – 1964)	physicist	AUSTRIA
Hess/ Walter (1881 – 1973)	physiologist	SWITZERLAND
Heymans/ Corneille Jean François (1892 – 1968)	physiologist	BELGIUM
Heyrovský/ Jaroslav (1890 – 1967)	chemist and inventor	CZECH REPUBLIC
Hilbert/ David (1862 – 1943)	mathematician	GERMANY
Hill/ Archibald Vivian (1886 – 1977)	physiologist	ENGLAND
Hinshelwood/ Cyril Norman (1897 – 1967)	physical chemist	ENGLAND
Hipparchus (ca 190 – 127 BC)	astronomer, geographer, and mathematician	GREECE
Hippocrates (c. 460 – c. 370 BC)	also known as Hippocrates II, was a Greek physician of the Age of Pericles (Classical Greece), and is considered one of the most outstanding figures in the history of medicine	Greece
Hodgkin/ Alan Lloyd (1914 – 1998)	physiologist and biophysicist	ENGLAND
Hodgkin/ Dorothy (1910 – 1994)	biochemist	ENGLAND
Hofer/ Bruno (1861 – 1916)	marine scientist and an environmentalist	GERMANY
Hoffmann/ Felix (1868 – 1946)	chemist, credited for the first synthesized medically useful forms of heroin and aspirin	GERMANY
Hooke/ Robert (1635 – 1703)	natural philosopher, architect and polymath	ENGLAND
Hopkins/ William (1793 – 1866)	mathematician and geologist	ENGLAND
Hörmander/ Lars Valter	mathematician who has	SWEDEN

Best European Scientist

scientists	profession	country
(1931 – 2012)	been called "the foremost contributor to the modern theory of linear partial differential equations	
Hornidge Porter/ George (1920 – 2002)	chemist	ENGLAND
Hurwicz/ Leonid (1917 – 2008)	economist and mathematician, He originated incentive compatibility and mechanism design	RUSSIA
Hutton/ James (1726 – 1797)	geologist, physician, chemical manufacturer, naturalist, and experimental agriculturalist	SCOTLAND
Huygens/ Christiaan (1629 – 1695)	mathematician and scientist	NETHERLANDS
Hypatia of Alexandria (AD 350 – 370; killed in 415)	Neoplatonist philosopher	GREECE
Ivanovich Lobachevsky Nikolai (1793 – 1856)	mathematician and geometer	RUSSIA
Jacob/ François (1920 – 2013)	biologist	FRANCE
Jacobi/ Carl Gustav Jacob (1804 – 1851)	mathematician	GERMANY
Janssen/ Zacharias (1585 – 1632)	spectacle-maker from Middelburg associated with the invention of the first optical telescope. Janssen is sometimes also credited for inventing the first truly compound microscope	NETHERLAND
Jenner/ Edward (1749 –	physician and scientist	ENGLAND

Best European Scientist

scientists	profession	country
1823)	who was the pioneer of smallpox vaccine,	
Jensen/ Johannes Hans Daniel (1907 – 1973)	nuclear physicist	GERMANY
Jerne/ Niels Kaj (1911 – 1994)	immunologist	DENMARK
Jordan/ Marie Ennemond Camille (1838 – 1922)	mathematician	FRANCE
Joule (Prescott)/ James (1818 – 1889)	physicist and brewer	ENGLAND
Junkers/ Hugo (1859 – 1935)	engineer and aircraft designer	GERMANY
Karrer/ Paul (1889 – 1971)	organic chemist	SWITZERLAND
Kastler/ Alfred (1902 – 1984)	physicist	FRANCE
Katz/ Bernard (1911 – 2003)	biophysicist	GERMANY
Kekulé/ Friedrich August (1829 – 1896)	organic chemist	GERMANY
Kepler/ Johannes (1571 – 1630)	mathematician, astronomer, and astrologer	GERMANY
Kirchoff/ Gustav (1824 – 1887)	physicist	GERMANY
Klein/ Christian Felix (1849 – 1925)	mathematician	GERMANY
Koch/ Robert (1843 – 1910)	physician and pioneering microbiologist, The founder of modern bacteriology	GERMANY
Kocher/ Emil Theodor (1841 – 1917)	physician and medical researcher	SWITZERLAND
Kohn/ Hedwig (1887 – 1964)	physicist	POLAND
Kolmogorov/ Andrey Nikolaevich (1903 – 1987)	mathematician	RUSSIA
Kraepelin/ Emil (1856 – 1926)	psychiatrist	GERMANY
Krebs/ Hans Adolf (1900 – 1981)	physician and biochemist	GERMANY
Kronecker/ Leopold (1823 –	mathematician	GERMANY

Best European Scientist

scientists	profession	country
1891)		
Kuhn/ Richard (1900 – 1967)	biochemist	AUSTRIA
Kummer/ Ernst Eduard (1810 – 1893)	mathematician	GERMANY
Kurt/ Alder (1902 – 1958)	chemist	GERMANY
Laennec/ Rene (1781 – 1826)	physician. He invented the stethoscope in 1816, while working at the Hôpital Necker and pioneered its use in diagnosing various chest conditions	FRANCE
Lagrange/ Joseph Louis (1736 – 1813)	mathematician, astronomer	ITALY
Lajos/ Janos (1903 – 1957)	mathematician, physicist, inventor	HUNGARY
Lamarck/ Jean Baptiste (1744 – 1829)	naturalist	FRANCE
Lambert/ Johann Heinrich (1727 – 1777)	mathematician, physicist, philosopher and astronomer	SWITZERLAND
Landsteiner/ Karl (1868 – 1943)	biologist and physician	AUSTRIA
Laplace/ Pierre Simon (1749 – 1827)	mathematician and astronomer	FRANCE
Laveran/ Charles Louis Alphonse (1845 – 1922)	physician	FRANCE
Lavoisier/ Antoine (1743 – 1794)	chemist	FRANCE
Lazarevich Ginzburg Vitaly (1916 – 2009)	theoretical physicist, astrophysicist	RUSSIA
Le Chatelier/ Henry Louis (1850 – 1936)	chemist. He is most famous for devising Le Châtelier's principle	FRANCE
Le jeune Dirichlet/ Johann Peter Gustav (1805 – 1859)	mathematician	GERMANY
Le Rond d'Alembert/ Jean (1717 – 1783)	mathematician, mechanician, physicist,	FRANCE

Best European Scientist

scientists	profession	country
	philosopher, and music theorist	
Lebesgue/ Henri Léon (1875 – 1941)	mathematician	FRANCE
Leclerc/ Georges Louis ; Comte de Buffon (1707 – 1788)	naturalist, mathematician, cosmologist, and encyclopedic author	FRANCE
Leibniz/ Gottfried Wilhelm (1646 – 1716)	mathematician and philosopher	GERMANY
Levi Civita/ Tullio (1873 – 1941)	mathematician	ITALY
Levi Montalcini/ Rita (1909 – 2012)	neurologist	ITALY
Levi Strauss Claude (1908 – 2009)	anthropologist and ethnologist	FRANCE
Lie/ Marius Sophus (1842 – 1899)	mathematician	NORWAY
Lilienthal/ Otto (1848 – 1896)	pioneer of aviation	GERMANY
Linnaeus/ Carl (1707 – 1778)	botanist, physician, and zoologist	SWEDEN
Liouville/ Joseph (1809 – 1882)	mathematician	FRANCE
Lipmann/ Fritz Albert (1899 – 1986)	biochemist	GERMANY
Lipperhey/ Hans (1570 – 1619)	inventor (telescope)	GERMANY
Lippmann/ Jonas Ferdinand Gabriel (1845 – 1921)	physicist and inventor	LUXEMBOURG
Lister/ Joseph (1827 – 1912)	surgeon and a pioneer of antiseptic surgery.	ENGLAND
Locke/ John (1632 – 1704)	philosopher and physician	ENGLAND
Loewi/ Otto (1873 – 1961)	pharmacologist	GERMANY
Logie Baird/ John (1888 – 1946)	engineer, innovator, inventor (television)	SCOTLAND
Lorentz/ Hendrik Antoon (1853 – 1928)	physicist	NETHERLANDS

Best European Scientist

scientists	profession	country
Lorenz/ Konrad (1903 – 1989)	zoologist, ethologist, and ornithologist	AUSTRIA
Lovelace/ Ada (1815 – 1852)	mathematician	ENGLAND
Lowry/ Thomas Martin (1874 – 1936)	physical chemist who developed the Brønsted–Lowry acid–base theory simultaneously with and independently of Johannes Nicolaus Brønsted and was a founder-member and president (1928–1930) of the Faraday Society	ENGLAND
Lucretius (99 BC – c. 55 BC)	philosopher	ITALY
Luria/ Salvador Edward (1912 – 1991)	microbiologist	ITALY
Lvovich Chebyshev/ Pafnuty (1821 – 1894)	mathematician	RUSSIA
Lwoff/ André Michel (1902 – 1994)	microbiologist	FRANCE
Lyell/ Charles (1797 – 1875)	geologist	SCOTLAND
Lynen/ Feodor Felix Konrad (1911 – 1979)	biochemist	GERMANY
Lysenko/ Trofim (1898 – 1976)	biologist and agronomist	RUSSIA
Mach/ Ernst (1838 – 1916)	physicist and philosopher	AUSTRIA
Mackintosh/ Charles (1766 – 1843)	chemist and inventor of waterproof fabrics. The Mackintosh raincoat	SCOTLAND
Maclaurin/ Colin (1698 – 1746)	mathematician	SCOTLAND
Macleod/ John James Rickard (1876 – 1935)	biochemist and physiologist	SCOTLAND
Maklai/ Nicholai (1846 – 1888)	anthropologist, ethnologist and a marine biologist who graduated from St. Petersburg University	RUSSIA
Malpighi/ Marcello (1628 –	physician and biologist	ITALY

Best European Scientist		
scientists	profession	country
1694)		
Marconi/ Guglielmo (1874 – 1937)	inventor and electrical engineer	ITALY
Markov/ Andrei Andreyevich (1856 – 1922)	mathematician	RUSSIA
Maxwell/ James Clerk (1831 – 1879)	mathematical physicist	SCOTLAND
Mayr/ Ernst (1904 – 2005)	evolutionary biologists, taxonomist, tropical explorer, ornithologist, and historian of science	GERMANY
Meade/ James Edward (1907 – 1995)	economist	ENGLAND
Mechnikov/ Ilya Ilyich (1845 – 1916)	biologist, zoologist and protozoologist	RUSSIA
Medawar/ Peter Brian (1915 – 1987)	biologist	ENGLAND
Megara/ Euclid (435 – c. 365 BCE)	philosopher	GREECE
Meitner/ Lise (1878 – 1968)	physicist	AUSTRIA
Mendel/ Gregor (1822 – 1884)	scientist (founder of the modern science of genetics)	CZECH REPUBLIC
Mendeleev/ Dmitri (1834 – 1907)	chemist and inventor	RUSSIA
Meucci/ Antonio (1808 – 1889)	inventor "voice-communication apparatus"	ITALY
Meyer/ Julius Lothar (1830 – 1895)	chemist, First person to draw the periodic table of chemical elements	GERMANY
Meyerhof/ Otto Fritz (1884 – 1951)	physician and biochemist	GERMANY
Millington Synge/ Richard Laurence (1914 - 1994)	biochemist	ENGLAND
Minkowski Hermann (1864 – 1909)	mathematician	GERMANY
Modigliani/ Franco (1918 –	economist	ITALY

Best European Scientist

scientists	profession	country
2003)		
Moissan/ Ferdinand Frederick Henri (1852 – 1907)	chemist	FRANCE
Monge/ Gaspard ; Comte de Péluse (1746 – 1818)	mathematician, the inventor of descriptive geometry	FRANCE
Monod/ Jacques Lucien(1910– 1976)	biologist	FRANCE
Moseley/ Henry (1887 – 1915)	physicist	ENGLAND
Mott/ Nevill Francis (1905 – 1996)	physicist	ENGLAND
Müller/ Paul Hermann (1899 – 1965)	chemist	SWITZERLAND
Müller von Königsberg `Regiomontanus' Johannes (1436 – 1476)	mathematician, astronomer, astrologer, translator, instrument maker	GERMANY
Murray/ John (1841 – 1914)	oceanographer and a marine biologist	SCOTLAND
Myrdal/ Karl Gunnar (1898 – 1987)	economist, sociologist	SWEDEN
Napier/ John (1550 – 1617)	mathematician, physicist, and astronomer.	SCOTLAND
Natta/ Giulio (1903 – 1979)	chemist	ITALY
Needham/ John (1713 – 1781)	biologist	ENGLAND
Nernst/ Walther Hermann (1864 – 1941)	physicist	GERMANY
Newton/ Isaac (1642 – 1727)	physicist and mathematician, natural philosopher	ENGLAND
Nicolle/ Charles Jules Henry (1866 – 1936)	bacteriologist	FRANCE
Niels/ Aage Bohr (1922 – 2009)	nuclear physicist	DENMARK
Niepce/ Nicephore (1765 –	inventor, now usually	FRANCE

scientists	profession	country
1833)	credited as the inventor of photography and a pioneer in that field	
Nikolayevich Semyonov/ Nikolay (1896 – 1986)	physicist and chemist	RUSSIA
Nobel/ Alfred (1833 – 1896)	chemist, engineer, innovator	SWEDEN
Noddack/ Ida (1896 – 1978)	chemist and physicist	GERMANY
Noether/ Amalie Emma (1882 – 1935)	mathematician	GERMANY
Nollet/ Jean (1700 – 1770)	clergyman and physicist	FRANCE
Oberth/ Hermann (1894 – 1989)	physicist and engineer	GERMANY
Ochoa de Albornoz/ Severo (1905 – 1993)	physician and biochemist	SPAIN
Oersted Hans/ Christian (1777 – 1851)	physicist and chemist	DENMARK
Ohlin Bertil Gotthard (1899 – 1979)	economist	SWEDEN
Ohm/ Georg (1789 – 1854)	physicist and mathematician	GERMANY
Onnes/ Heike Kamerlingh (1853 – 1926)	physicist	NETHERLANDS
Onsager/ Lars (1903 – 1976)	physical chemist and theoretical physicist	NORWAY
Oresme/ Nicole (ca 1322 – 1382)	philosopher (He wrote influential works on economics, mathematics, physics, astrology and astronomy, philosophy, and theology)	FRANCE
Ostwald/ Wilhelm (1853 – 1932)	chemist	GERMANY
Paget Thomson/ George (1892 – 1975)	physicist	ENGLAND
Palade/ George Emil (1912 – 2008)	cell biologist	ROMANIA
Paracelsus (1493 – 1541)	physician, botanist,	SWITZERLAND

Best European Scientist

scientists	profession	country
	alchemist, astrologer, and general occultist	
Pascal/ Blaise (1623 – 1662)	mathematician, physicist, inventor	FRANCE
Pasteur/ Louis (1822 – 1895)	chemist and microbiologist	FRANCE
Pavlov/ Ivan (1849 – 1936)	physiologist	RUSSIA
Peano/ Giuseppe (1858 – 1932)	mathematician	ITALY
Perkin/ William Henry (1838 – 1907)	chemist best known for his accidental discovery, at the age of 18, of the first aniline dye, mauveine.	ENGLAND
Perrin Jean Baptiste (1870 – 1942)	physicist	FRANCE
Perutz/ Max Ferdinand (1914 – 2002)	molecular biologist	AUSTRIA
Petrovic Alas/ Mihailo (1868 – 1943)	mathematician and inventor	SERBIA
Pfleumer/ Fritz (1881 – 1945)	engineer who invented magnetic tape for recording sound	AUSTRIA
Piaget/ Jean (1896 – 1980)	psychologist and philosopher	SWITZERLAND
Planck/ Max (1858 – 1947)	theoretical physicist	GERMANY
Plato (424/423 – 348/347 BCE)	he was a philosopher in Classical Greece and the founder of the Academy in Athens, the first institution of higher learning in the Western world	Greece
Plücker/ Julius (1801 – 1868)	mathematician and physicist	GERMANY
Poincaré/ Jules Henri (1854 – 1912)	mathematician, theoretical physicist,	FRANCE

Best European Scientist

scientists	profession	country
	engineer, and a philosopher of science	
Poisson/ Siméon Denis (1781 – 1840)	mathematician, geometer, and physicist	FRANCE
Poncelet/ Jean Victor (1788 – 1867)	engineer and mathematician	FRANCE
Porter/ Rodney Robert (1917 – 1985)	biochemist	ENGLAND
Porter Martin Archer John (1910 – 2002)	chemist	ENGLAND
Potter/ Beatrix (1866 – 1943)	illustrator, natural scientist and conservationist	ENGLAND
Powell/ Cecil Frank (1903 – 1969)	physicist	ENGLAND
Pregl/ Fritz (1869 – 1930)	chemist and physician	SLOVENIA
Prelog/ Vladimir (1906 – 1998)	organic chemist	CROATIA
Priestley/ Joseph (1733 – 1804)	theologian, Dissenting clergyman, natural philosopher	ENGLAND
Prigogine/ Ilya Romanovich (1917 – 2003)	physical chemist	RUSSIA
Prokhorov / Alexander Mikhaylovich (1916 – 2002)	physicist	RUSSIA
Ptolemy/ Claudius (c. AD 90 – c.168)	astronomer and mathematician, Ptolemy's geocentric views on the structure of the universe dominated astronomy until the Scientific Revolution	GREECE
Pythagoras of Samos (c. 570 BC – c. 495 BC)	philosopher, mathematician, and founder of the religious movement called Pythagoreanism	GREECE
Ramsay/ William (1852 – 1916)	chemist	SCOTLAND

Best European Scientist

scientists	profession	country
Raoult/ François (1830 – 1901)	chemist who conducted research into the behavior of solutions, especially their physical propertie	FRANCE
Ray/ John (1627 – 1705)	naturalist	ENGLAND
Redi/ Francesco (1626 – 1697)	physician, naturalist	ITALY
Rees Wilson/ Charles Thomson (1869 – 1959)	physicist and meteorologist	SCOTLAND
Reichstein/ Tadeusz (1897 – 1996)	chemist	POLAND
Reis/ Philipp (1834 – 1874)	scientist and inventor	GERMANY
Richet/ Charles Robert (1850 – 1935)	physiologist	FRANCE
Riemann/ Georg Friedrich Bernhard (1826 – 1866)	mathematician	GERMANY
Ritter von Frisch/ Karl (1886 – 1982)	ethologist	AUSTRIA
Robertus Todd/ Alexander (1907 – 1997)	biochemist	SCOTLAND
Robinson/ Robert (1886 – 1975)	organic chemist	ENGLAND
Roentgen/ Wilhelm Conrad (1845 – 1923)	physicist	GERMANY
Rohrer/ Heinrich (1933 – 2013)	physicist	SWITZERLAND
Rorschach Hermann (1884 – 1922)	psychiatrist and psychoanalyst,	SWITZERLAND
Ross/ Ronald (1857 – 1932)	medical doctor	ENGLAND
Rutherford/ Daniel (1749 – 1819)	physician, chemist and botanist who is most famous for the isolation of nitrogen in 1772	SCOTLAND
Rutherford/ Ernest (1871 – 1937)	physicist	ENGLAND

Best European Scientist

scientists	profession	country
Ružička/ Leopold (1887 – 1976)	scientist, chemist	CROATIA
Ryberg Finsen/ Niels (1860 – 1904)	physician and scientist	FAROE ISLANDS
Sabatier/ Paul (1854 – 1941)	chemist	FRANCE
Sanger/ Frederick (1918 – 2013)	biochemist	ENGLAND
Santorio/ Santorio (1561 – 1636)	physiologist, physician, and professor. He introduced the quantitative approach into medicine and, as his pupil, introduced the mechanistic principles of Galileo Galilei to medicine	ITALY
Sars/ Georg (1837 – 1927)	marine taxonomist. Sars discovered a new species of mysids and ostracods. His fileds of research included various branches of fishery.	NORWAY
Sax/ Adolphe (1814 – 1894)	musical instrument designer	BELGIUM
Scheele/ Carl Wilhelm (1742 – 1786)	pharmaceutical chemist	SWEDEN
Schottky/ Walter (1886 – 1976)	physicist (who played a major early role in developing the theory of electron and ion emission phenomena)	GERMANY
Schrödingers/ Erwin (1887 – 1961)	physicist	AUSTRIA
Schwann/ Theodor (1810 – 1882)	physiologist	GERMANY
Segrè/ Emilio Gino (1905 – 1989)	physicist	ITALY
Selye/ Hans (1907 – 1982)	endocrinologist	AUSTRIA

Best European Scientist

scientists	profession	country
Sherrington/ Charles Scott (1857 – 1952)	neurophysiologist, histologist, bacteriologist, and a pathologist	ENGLAND
Siegbahn/ Karl Manne Georg (1886 – 1978)	physicist	SWEDEN
Siegel/ Carl Ludwig (1896 – 1981)	mathematician	GERMANY
Siemens/ Ernst Werner (1816 – 1892)	inventor and industrialist	GERMANY
Sikorsky/ Igor (1889 – 1972)	aviation pioneer in both helicopters and fixed-wing aircraft	RUSSIA
Sinton Walton/ Ernest Thomas (1903 – 1995)	physicist	IRELAND
Smith/ Henry John Stephen (1826 – 1883)	mathematician	IRELAND
Smith/ William (1769 – 1839)	geologist	ENGLAND
Soddy/ Frederick (1877 – 1956)	radiochemist	ENGLAND
Somerville/ Mary (1780 – 1872)	science writer and polymath, at a time when women's participation in science was discouraged. She studied mathematics and astronomy	SCOTLAND
Sommerfeld/ Arnold (1868 – 1951)	theoretical physicist	GERMANY
Spemann/ Hans (1869 – 1941)	embryologist	GERMANY
Stagira/ Aristotle (384 – 322 BC)	philosopher	MACEDONIA
Stark/ Johannes (1874 – 1957)	physicist	GERMANY
Staudinger/Hermann (1881–1965)	chemist	GERMANY
Steiner/ Jakob (1796 – 1863)	mathematician	SWITZERLAND
Stern/ Otto (1888 – 1969)	physicist	GERMANY

Best European Scientist

scientists	profession	country
Stevin/ Simon (1549 – 1620)	mathematician and military engineer	BELGIUM
Stone/ John Richard Nicholas (1913 – 1991)	economist	ENGLAND
Strutt/ John William, 3rd Baron Rayleigh (1842 – 1919)	physicist	ENGLAND
Svedberg/ Theodor (1884 – 1971)	chemist	SWEDEN
Swainson/ William John (1789 – 1855)	ornithologist, malacologist, conchologist, entomologist and artist	ENGLAND
Sylvester/ James Joseph (1814 – 1897)	mathematician	ENGLAND
Sylvius/ Franciscus (1614 – 1672)	chemical biologists, introduced the idea of chemical affinity to explain the human body's use of salts. He and his followers contributed greatly to the study of digestion and body fluids	NETHERLANDS
Syracuse/ Archimedes (287 – 212 BC)	mathematician, physicist, engineer, inventor, and astronomer	GREECE
Szilard/ Leo (1898 – 1964)	physicist and inventor	HUNGARY
Tamm/ Igor Yevgenyevich (1895 – 1971)	physicist	RUSSIA
Tarentum/ Archytas (ca 428 – 347 BC)	philosopher, mathematician, astronomer, statesman, and strategist	GREECE
Taylor/ Brook (1685 – 1731)	mathematician	ENGLAND
Telkes/ Maria (1900 – 1995)	scientist and inventor (solar energy technologies)	HUNGARY

Best European Scientist

scientists	profession	country
Teller/ Edward (1908 – 2003)	theoretical physicist	HUNGARY
Terentius Varro/ Marcus (116 BC – 27 BC)	he was an ancient Roman scholar and writer.	Italy
Tesla/ Nikola (1856 – 1943)	inventor, electrical engineer, mechanical engineer, and futurist	SERBIA
Thales of Miletus (ca 624 – 546 BC)	philosopher	GREECE
Thompson/ William (1805 – 1852)	naturalist celebrated for his founding studies of the natural history of Ireland, especially in ornithology and marine biology	IRELAND
Thomson/ Joseph John (1856 – 1940)	physicist	ENGLAND
Thomson/ William (1824 – 1907)	mathematical physicist and engineer	IRELAND
Tinbergen/ Jan (1903 – 1994)	economist	NETHERLANDS
Tinbergen/ Nikolaas (1907 – 1988)	biologist and ornithologist	NETHERLANDS
Tiselius/ Arne Wilhelm Kaurin (1902 – 1971)	biochemist	SWEDEN
Torricelli/ Evangelista (1608 – 1647)	physicist and mathematician	ITALY
Turing/ Alan Mathison (1912 – 1954)	mathematician, logician, cryptanalyst, philosopher, computer scientist, mathematical biologist	ENGLAND
Van der Waals/ Johannes Diderik (1837 – 1923)	theoretical physicist and thermodynamicist	NETHERLANDS
Van Helmont/ Jan Baptist (1580 – 1644)	chemist, physiologist, and physician	BELGIUM
Van Leeuwenhoek/ Antonie (1632 – 1723)	scientist "the Father of Microbiology"	NETHERLANDS

Best European Scientist

scientists	profession	country
Van 't Hoff, Jr./ Jacobus Henricus (1852 – 1911)	physical and organic chemist	NETHERLANDS
Vane/ John Robert (1927 – 2004)	pharmacologist	ENGLAND
Vasilyevna Kovalevskaya/ Sofia (1850 – 1891)	mathematician	RUSSIA
Venn/John (1834 – 1923)	logician and philosopher noted for introducing the Venn diagram, used in the fields of set theory, probability, logic, statistics, and computer science	ENGLAND
Verbiest/ Ferdinand (1623 – 1688)	mathematician and astronomer	BELGIUM
Vernadsky/ Vladimir (1863 – 1945)	mineralogist and geochemist	RUSSIA
Vesalius/ Andreas (1514 – 1564)	anatomist, physician	BELGIUM
Viète/ François (1540 – 1603)	mathematician	FRANCE
Virchow/ Rudolf (1821 – 1902)	doctor, anthropologist, pathologist, prehistorian, biologist, writer, editor	GERMANY
Virtanen/ Artturi (1895–1973)	chemist	FINLAND
Viscount de Duve/ Christian René (1917 – 2013)	cytologist and biochemist	BELGIUM
Volta/ Alessandro (1745 – 1827)	physicist	ITALY
Von Ardenne/ Manfred (1907 – 1997)	physicist and inventor	GERMANY
Von Baer Karl/ Ernst (1792 – 1876)	naturalist, biologist, geologist, meteorologist, geographer	ESTONIA
Von Braun/ Wernher (1912 – 1977)	aerospace engineer and space architect	GERMANY

Best European Scientist

scientists	profession	country
Von Drais/ Karl Friedrich (1785 – 1851)	inventor, who invented the Laufmaschine ("running machine"), also later called the velocipede,	GERMANY
Von Euler/ Ulf Svante (1905 – 1983)	physiologist and pharmacologist	SWEDEN
Von Euler Chelpin/ Hans Karl August Simon (1873 – 1964)	biochemist	GERMANY
Von Fraunhofer/ Joseph (1787 – 1826)	optician	GERMANY
Von Guericke/ Otto (1602 – 1686)	scientist, inventor the air pump, and did the first experiments with vacuums.	GERMANY
Von Haller/ Albrecht (1708 – 1777)	anatomist, physiologist, naturalist	SWITZERLAND
Von Helmholtz / Hermann (1821 – 1894)	physician and physicist	GERMANY
Von Humboldt / Alexander (1769 – 1859)	geographer, naturalist, and explorer	GERMANY
Von Laue/ Max (1879 – 1960)	physicist	GERMANY
Von Liebig/ Justus (1803 – 1873)	chemist	GERMANY
Von Mayer/ Julius Robert (1814 – 1878)	physician and physicist and one of the founders of thermodynamics	GERMANY
Von Mohl/ Hugo (1805 – 1872)	botanist	GERMANY
Von Ohain/ Hans (1911 – 1998)	engineer, and designer of the first operational jet engine	GERMANY
Wagner Jauregg/ Julius (1857 – 1940)	physician	AUSTRIA
Wallach/ Otto (1847 – 1931)	chemist	GERMANY
Wallis/ John (1616 – 1703)	mathematician	ENGLAND

Best European Scientist

scientists	profession	country
Warburg/ Otto Heinrich (1883 – 1970)	physiologist, medical doctor	GERMANY
Watt/ James (1736 – 1819)	inventor and mechanical engineer	SCOTLAND
Wegener/ Alfred (1880 – 1930)	polar researcher, geophysicist and meteorologist	GERMANY
Weierstrass/ Karl Wilhelm Theodor (1815 – 1897)	mathematician	GERMANY
Weil/ André (1906 – 1998)	influential mathematician of the 20th century	FRANCE
Weißkopf/ Gustav (1874 – 1927)	engineer (designed and built gliders, flying machines and engines)	GERMANY
Werner/ Alfred (1866 – 1919)	chemist	SWITZERLAND
Weyl/ Hermann Klaus Hugo (1885 – 1955)	mathematician, theoretical physicist and philosopher	GERMANY
Whyte Black/ James (1924 – 2010)	pharmacologist	SCOTLAND
Wieland/ Heinrich Otto (1877 – 1957)	chemist	GERMANY
Wien/ Wilhelm Carl Werner Otto Fritz Franz (1864 – 1928)	physicist	GERMANY
Wigner/ Eugene Paul (1902 – 1995)	theoretical physicist and mathematician	HUNGARY
Wilkins/ Maurice Hugh Frederick (1916 – 2004)	physicist and molecular biologist	ENGLAND
Wilkinson/ Geoffrey (1921 – 1996)	chemist	ENGLAND
Willans Richardson/ Owen (1879 – 1959)	physicist	ENGLAND

Best European Scientist

scientists	profession	country
Willis/ Thomas (1621 – 1675)	doctor (who played an important part in the history of anatomy, neurology and psychiatry)	ENGLAND
Willstätter/ Richard Martin (1872 – 1942)	organic chemist	GERMANY
Windaus/ Adolf Otto Reinhold (1876 – 1959)	chemist	GERMANY
Wittig/ Georg (1897 – 1987)	chemist	GERMANY
Wöhler/ Friedrich (1800 – 1882)	chemist	GERMANY
Wreyford Norrish/ Ronald George (1897 – 1978)	chemist	ENGLAND
Wundt/ Wilhelm (1832 – 1920)	physician, physiologist, philosopher, and professor	GERMANY
Wyville Thomson/ Charles (1830 – 1882)	marine biologist who was the chief scientist on the Challenger Expedition.	SCOTLAND
Zeeman/ Pieter (1865 – 1943)	physicist	NETHERLANDS
Zernike/ Frits (1888 – 1966)	physicist	NETHERLANDS
Ziegler/ Karl Waldemar (1898 – 1973)	chemist	GERMANY
Zsigmondy/ Richard Adolf (1865 – 1929)	chemist	AUSTRIA
Zuse/ Konrad (1910 – 1995)	engineer, inventor and computer pioneer	GERMANY

Arya Bàhram Picture Collection
Aage Niels Bohr (1922 – 2009)   nuclear physicist
DENMARK

Arya Bàhram Picture Collection
Abram Besicovitch (1891 – 1970) mathematician
RUSSIA

Arya Bàhram Picture Collection
Ada Lovelace (1815 – 1852) mathematician
ENGLAND

Arya Bàhram Picture Collection
Adalbert Czerny (1863 – 1941) pediatrician
AUSTRIA

Arya Bàhram Picture Collection
Adolf Friedrich Johann Butenandt (1903 – 1995) biochemist Germany

Arya Bàhram Picture Collection
Adolf Otto Reinhold Windaus (1876 – 1959) chemist
GERMANY

Arya Bàhram Picture Collection Adolphe Sax (1814 – 1894) musical instrument designer BELGIUM

GALENVS

Arya Bàhram Picture Collection
Aelius Galenus (AD 129 – c.200/c.216) surgeon and philosopher, physician
GREECE

Arya Bàhram Picture Collection
Agnes Arber (1879 – 1960) plant morphologist and anatomist, historian of botany and philosopher of biology
ENGLAND

Arya Bàhram Picture Collection
Alan Lloyd Hodgkin (1914 – 1998) physiologist and biophysicist
ENGLAND

Arya Bàhram Picture Collection
Alan Mathison Turing
(1912 – 1954)
mathematician, logician, cryptanalyst, philosopher, computer scientist, mathematical biologist   ENGLAND

Arya Bàhram Picture Collection
Albert Einstein (1879 – 1955) theoretical physicist and philosopher of science
GERMANY

Arya Bàhram Picture Collection
Albert Szent Györgyi
(1893 – 1986)
physiologist
HUNGARY

Arya Bàhram Picture Collection
Albrecht von Haller (1708 – 1777) anatomist, physiologist, naturalist   SWITZERLAND

Arya Bàhram Picture Collection
Aleksandr Danilovich Aleksandrov (1912 – 1999)
mathematician, physicist, philosopher and mountaineer.
RUSSIA

Arya Bàhram Picture Collection
Alessandro Volta
(1745 – 1827) physicist
ITALY

Arya Bàhram Picture Collection
Alexander Bain (1818 – 1903) philosopher and educationalist
SCOTLAND

Arya Bàhram Picture Collection
Alexander Fleming (1881 – 1955) biologist, pharmacologist and botanist   SCOTLAND

Arya Bàhram Picture Collection
Alexander Graham Bell (1847 – 1922) scientist, inventor, engineer and innovator
SCOTLAND

Arya Bàhram Picture Collection

Alexander Mikhaylovich Prokhorov (1916 – 2002) physicist RUSSIA

Arya Bàhram Picture Collection
Alexander Robertus Todd (1907 – 1997) biochemist
SCOTLAND

Arya Bahram Picture Collection
Alexander Von Humboldt (1769 – 1859) geographer, naturalist, and explorer
GERMANY

Arya Bàhram Picture Collection
Alexandre Brongniart (1770 – 1847) chemist, mineralogist, and zoologist   FRANCE

Arya Bàhram Picture Collection

Alexis Carrel (1873 – 1944) surgeon and biologist FRANCE

Arya Bàhram Picture Collection
Alexis Claude Clairaut (1713 – 1765) mathematician, astronomer, geophysicist   FRANCE

Arya Bàhram Picture Collection
Alfred Binet (1857 – 1911) psychologist
FRANCE

Arya Bàhram Picture Collection
Alfred Kastler (1902 – 1984) physicist
FRANCE

Arya Bàhram Picture Collection
Alfred Nobel (1833 – 1896) chemist, engineer, innovator
SWEDEN

Arya Bàhram Picture Collection
Alfred Wegener (1880 – 1930) polar researcher, geophysicist and meteorologist GERMANY

Arya Bàhram Picture Collection
Alfred Werner (1866 – 1919) chemist
SWITZERLAND

Arya Bàhram Picture Collection
Alister Clavering Hardy (1896 – 1985) marine biologist who was an expert on marine ecosystems and zooplankton
ENGLAND

Arya Bàhram Picture Collection
Allvar Gullstrand (1862 – 1930) ophthalmologist and optician   SWEDEN

Arya Bàhram Picture Collection
Amalie Emma Noether
(1882 – 1935)
mathematician
GERMANY

Arya Bàhram Picture Collection
Anders Celsius (1701 – 1744) astronomer, physicist and mathematician SWEDEN

Arya Bàhram Picture Collection
Andre Marie Ampère (1775 – 1836) physicist and mathematician   FRANCE

Arya Bàhram Picture Collection
André Michel Lwoff (1902 – 1994) microbiologist
FRANCE

Arya Bàhram Picture Collection
André Weil (1906 – 1998) influential mathematician of the 20th century
FRANCE

Arya Bàhram Picture Collection
Andreas Vesalius
(1514 – 1564)
anatomist, physician
BELGIUM

Arya Bàhram Picture Collection
Andrei Andreyevich Markov (1856 – 1922)
mathematician
RUSSIA

Arya Bàhram Picture Collection
Andrew Fielding Huxley (1917 – 2012) physiologist and biophysicist   ENGLAND

Arya Bàhram Picture Collection
Andrey Nikolaevich Kolmogorov (1903 – 1987) mathematician RUSSIA

Arya Bàhram Picture Collection
Antoine Lavoisier
(1743 – 1794) chemlst
FRANCE

Arya Bàhram Picture Collection
Anton Dohrn (1840 – 1909) marine biologist, He had mastered not only medicine but also zoology
GERMANY

Arya Bàhram Picture Collection
Antonie van Leeuwenhoek (1632 – 1723) scientist "the Father of Microbiology"
NETHERLANDS

Arya Bàhram Picture Collection
António Caetano de Abreu Freire Egas Moniz (1874 – 1955) neurologist and the developer of cerebral angiography PORTUGAL

Arya Bàhram Picture Collection
Antonio Luigi Cremona (1830 – 1903) mathematician
ITALY

Arya Bàhram Picture Collection
Antonio Meucci (1808 – 1889)   inventor of voice-communication apparatus
ITALY

**Arya Bàhram Picture Collection
Archer John Porter Martin (1910 – 2002) chemist
ENGLAND**

Arya Bàhram Picture Collection
Archibald Vivian Hill (1886 – 1977) physiologist
ENGLAND

Arya Bàhram Picture Collection
Archimedes of Syracuse (287 – 212 BC) mathematician, physicist, engineer, inventor, astronomer
GREECE

Arya Bahram Picture Collection
Archytas of Tarentum (ca 428 – 347 BC) philosopher, mathematician, astronomer, statesman, strategist
GREECE

Arya Bàhram Picture Collection
Aristotle of Stagira (384 – 322 BC) philosopher
MACEDONIA

Arya Bàhram Picture Collection
Arne Wilhelm Kaurin Tiselius (1902 – 1971) biochemist
SWEDEN

Arya Bàhram Picture Collection
Arnold Sommerfeld
(1868 – 1951)
theoretical physicist
GERMANY

Arya Bàhram Picture Collection
Arthur Cayley (1821-1895) mathematician
ENGLAND

Arya Bàhram Picture Collection
Arthur Eddington (1882 – 1944) astronomer, physicist, and mathematician ENGLAND

Arya Bàhram Picture Collection
Arthur Harden (1865 – 1940) biochemist
ENGLAND

Arya Bàhram Picture Collection
Artturi Virtanen (1895–1973) chemist
FINLAND

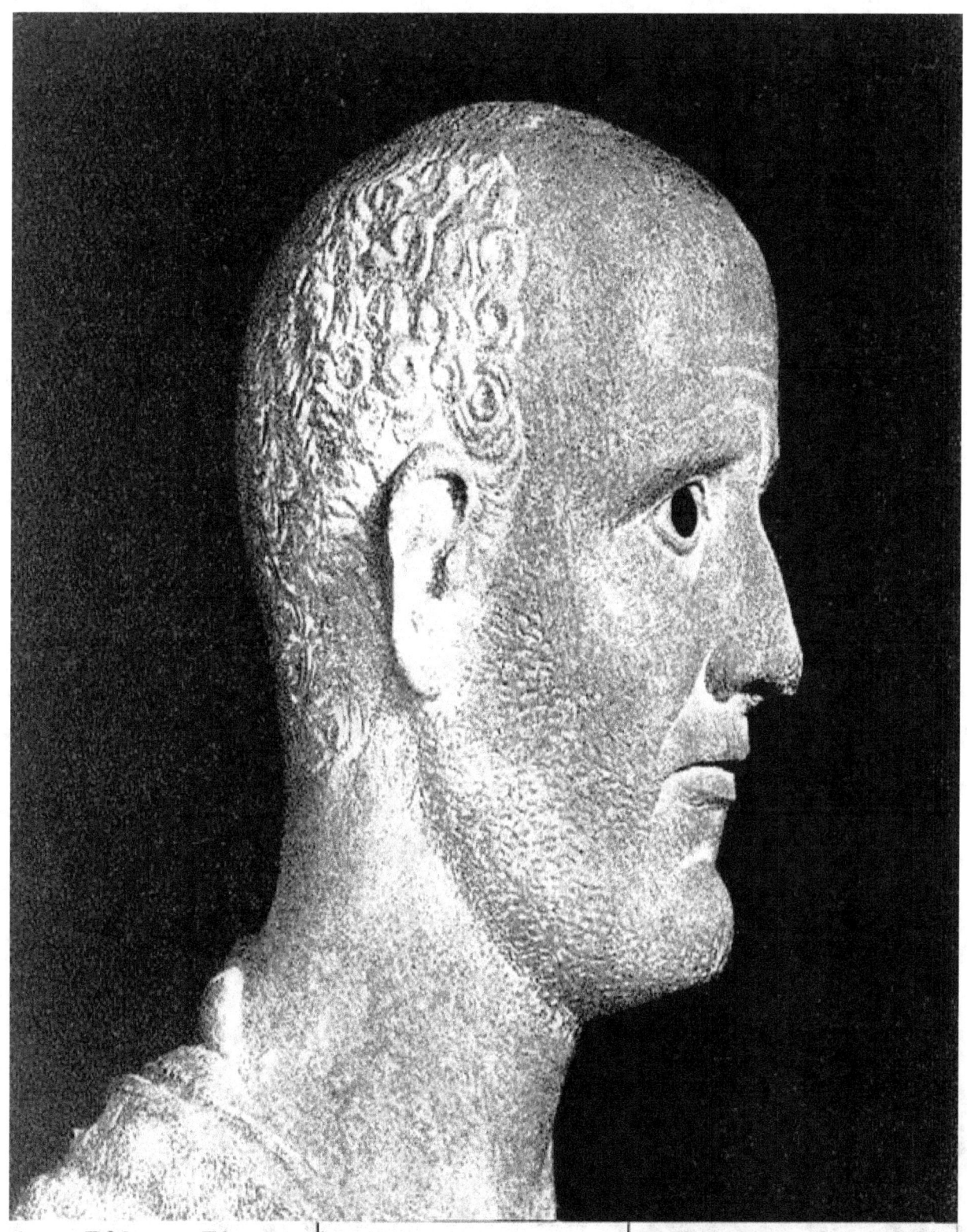

Arya Bàhram Picture Collection

Asclepiades : (c. 124 or 129 – 40 BC) GREECE, was a Greek physician born at Prusa in Bithynia in Asia Minor flourished at Rome, where he established Greek medicine near the end of the 2nd century BC. He attempted to build a new theory of disease

Arya Bàhram Picture Collection
Augustin Cauchy (1789 – 1857) mathematician
FRANCE

Arya Bàhram Picture Collection Aulus Cornelius Celsus (c. 25 BC – c. 50 AD) GREECE, he was a Roman encyclopaedist, known for his extant medical work, De Medicina, which is believed to be the only surviving section of a much larger encyclopedia.

Arya Bàhram Picture Collection
Bartolomeo Cristofori (1655 – 1731) maker of musical instruments, generally regarded as the inventor of the piano ITALY

Arya Bàhram Picture Collection
Beatrix Potter (1866 – 1943) illustrator, natural scientist and conservationist
ENGLAND

Arya Bàhram Picture Collection
Bernard Katz (1911 – 2003) biophysicist
GERMANY

Arya Bàhram Picture Collection
Bertil Gotthard Ohlin
(1899 – 1979)
economist
SWEDEN

Arya Bàhram Picture Collection
Blaise Pascal (1623 – 1662) mathematician, physicist, inventor FRANCE

Arya Bàhram Picture Collection
Brook Taylor (1685 – 1731) mathematician
ENGLAND

Arya Bàhram Picture Collection
Bruno Hofer (1861 – 1916) marine scientist and an environmentalist   GERMANY

Arya Bàhram Picture Collection
Camillo Golgi (1843 – 1926) physician, pathologist, scientist ITALY

Arya Bàhram Picture Collection
Carl Benz (1844 – 1929) engine designer and car engineer, inventor GERMANY

Arya Bàhram Picture Collection
Carl Bjerknes (1825 – 1903) mathematician and physicist
NORWAY

Arya Bàhram Picture Collection
Carl Bosch (1874 – 1940) chemist and engineer
GERMANY

Arya Bàhram Picture Collection
Carl Chun (1852 – 1914)   marine biologist
GERMANY

Arya Bàhram Picture Collection
Carl Ferdinand Cori (1896 – 1984) biochemist and pharmacologist   CZECH REPUBLIC

Arya Bàhram Picture Collection
Carl Friedrich Gauss (1777 – 1855) mathematician
GERMANY

Arya Bàhram Picture Collection
Carl Gustav Jacob Jacobi (1804 – 1851) mathematician
GERMANY

Arya Bàhram Picture Collection
Carl Linnaeus (1707 – 1778) botanist, physician, and zoologist SWEDEN

Arya Bàhram Picture Collection
Carl Ludwig Siegel
(1896 – 1981)
mathematician
GERMANY

Arya Bahram Picture Collection
Carl Wilhelm Scheele (1742 – 1786) pharmaceutical chemist SWEDEN

Arya Bàhram Picture Collection
Charles Augustin de Coulomb (1736 – 1806) physicist
FRANCE

Arya Bàhram Picture Collection
Charles Babbage (1791 – 1871) mathematician,
philosopher, inventor, mechanical engineer
ENGLAND

Arya Bàhram Picture Collection
Charles Darwin (1809 – 1882)   naturalist and geologist
ENGLAND

Arya Bàhram Picture Collection
Charles Édouard Guillaume (1861 – 1938) physicist
SWITZERLAND

Arya Bàhram Picture Collection
Charles Glover Barkla (1877 – 1944) physicist
ENGLAND

Arya Bàhram Picture Collection
Charles Hermite (1822 – 1901)   mathematician
FRANCE

Arya Bàhram Picture Collection
Charles Jules Henry Nicolle (1866 – 1936) bacteriologist
FRANCE

Arya Bàhram Picture Collection
Charles Louis Alphonse Laveran (1845 – 1922) physician
FRANCE

Arya Bàhram Picture Collection
Charles Lyell (1797 – 1875) geologist
SCOTLAND

Arya Bàhram Picture Collection
Charles Mackintosh (1766 – 1843) chemist and inventor of waterproof fabrics. The Mackintosh raincoat
SCOTLAND

Arya Bàhram Picture Collection
Charles Robert Richet (1850 – 1935) physiologist
FRANCE

Arya Bàhram Picture Collection
Charles Scott Sherrington (1857 – 1952)
neurophysiologist, histologist, bacteriologist, and a pathologist   ENGLAND

Arya Bàhram Picture Collection
Charles Thomson Rees Wilson (1869 – 1959) physicist and meteorologist SCOTLAND

Arya Bàhram Picture Collection
Charles Wyville Thompson (1830 – 1882) marine biologist who was the chief scientist on the Challenger Expedition.
SCOTLAND

Arya Bàhram Picture Collection
Christiaan Eijkman (1858 – 1930) physician and professor of physiology NETHERLANDS

Arya Bàhram Picture Collection
Christiaan Huygens (1629 – 1695) mathematician and scientist NETHERLANDS

Arya Bàhram Picture Collection
Christian Felix Klein (1849 – 1925) mathematician
GERMANY

Arya Bàhram Picture Collection
Christian René, viscount de Duve (1917 – 2013) cytologist and biochemist
BELGIUM

Arya Bàhram Picture Collection
Chrysippus (c. 279 BC – c. 206 BC) philosopher
GREECE

Arya Bàhram Picture Collection
Claude Bachet (1581 – 1638)   mathematician, linguist
FRANCE

Arya Bàhram Picture Collection
Claude Bernard (1813 – 1878) physiologist
FRANCE

Arya Bàhram Picture Collection
Claude Levi Strauss (1908 – 2009) anthropologist and ethnologist FRANCE

Arya Bàhram Picture Collection
Claudius Ptolemy (c. AD 90 – c.168) astronomer and mathematician, Ptolemy's geocentric views on the structure of the universe dominated astronomy until the Scientific Revolution GREECE

Arya Bàhram Picture Collection

Clive William John Granger (4 September 1934 – 27 May 2009) was a British economist

Arya Bàhram Picture Collection
Colin Maclaurin (1698 – 1746) mathematician
SCOTLAND

Arya Bàhram Picture Collection
Corneille Jean François Heymans (1892 – 1968)
physiologist BELGIUM

Arya Bàhram Picture Collection
Cyril Norman Hinshelwood (1897 – 1967) physical chemist ENGLAND

Arya Bàhram Picture Collection

Daniel Bernoulli (1700 – 1782) mathematician and physicist SWITZERLAND

Arya Bàhram Picture Collection
Daniel Bovet (1907 – 1992) pharmacologist
SWITZERLAND

Arya Bàhram Picture Collection
Daniel Rutherford (1749 – 1819) Daniel Rutherford was a Scottish botanist, chemist, and a physician. His most remarkable work is the discovery of Nitrogen
SCOTLAND

Arya Bàhram Picture Collection
David Hilbert (1862 – 1943) mathematician
GERMANY

Arya Bàhram Picture Collection

Democritus (c. 460 – c. 370 BC) influential pre-Socratic philosopher GREECE

Arya Bahram Picture Collection
Dennis Gabor (1900 – 1979) electrical engineer and physicist HUNGARY

Arya Bàhram Picture Collection
Diophantus of Alexandria (AD 201 – AD 285)
mathematician   GREECE

Arya Bàhram Picture Collection
Dmitri Mendeleev (1834 – 1907) chemist and inventor
RUSSIA

Arya Bàhram Picture Collection
Dorothy Hodgkin (1910 – 1994) biochemist
ENGLAND

Arya Bàhram Picture Collection
Edgar Douglas Adrian (1889 – 1977) electrophysiologist
ENGLAND

Arya Bàhram Picture Collection
Edmund Halley (1656 – 1742) astronomer, geophysicist, mathematician, meteorologist, and physicist
ENGLAND

Arya Bàhram Picture Collection
Eduard Buchner (1860 – 1917)   chemist and zymologist
GERMANY

Arya Bàhram Picture Collection
Edward Jenner (1749 – 1823) physician and scientist who was the pioneer of smallpox vaccine
ENGLAND

Arya Bahram Picture Collection
Edward Teller (1908 – 2003) theoretical physicist
HUNGARY

Arya Bàhram Picture Collection
Élie Cartan (1869 – 1951) mathematician
FRANCE

Arya Bàhram Picture Collection
Elizabeth Blackwell (1821 – 1910) doctor
ENGLAND

Arya Bàhram Picture Collection
Emil Fischer (1852 – 1919) chemist
GERMANY

Arya Bàhram Picture Collection Emil Kraepelin (1856 – 1926) psychiatrist GERMANY

Arya Bàhram Picture Collection
Emil Theodor Kocher (1841 – 1917) physician and medical researcher SWITZERLAND

Arya Bàhram Picture Collection
Emile Berliner (1851 – 1929) inventor "disc record gramophone" GERMANY

Arya Bàhram Picture Collection
Émilie du Châtelet (1706 – 1749) mathematician, physicist FRANCE

Arya Bàhram Picture Collection
Emilio Gino Segrè (1905 – 1989) physicist
ITALY

Arya Bàhram Picture Collection
Enrico Fermi (1901 – 1954) physicist
ITALY

Arya Bàhram Picture Collection
Eratosthenes (c. 276 BC – c. 195/194 BC) mathematician, geographer, poet, astronomer, and music theorist
GREECE

Arya Bàhram Picture Collection

Ernest Rutherford, (* 30.August 1871 in SpringGrove bei Nelson ;† 19.Oktober 1937 in Cambridge, Vereinigtes Königreich) war ein Physiker, der 1908 den Nobelpreis für Chemie erhielt.

Arya Bàhram Picture Collection
Ernest Thomas Sinton Walton (1903 – 1995) physicist
IRELAND

Arya Bàhram Picture Collection
Ernst Boris Chain (1906 – 1979) biochemist
GERMANY

Arya Bàhram Picture Collection
Ernst Eduard Kummer (1810 – 1893) mathematician
GERMANY

Arya Bàhram Picture Collection
Ernst Haeckel (1834 – 1919) biologist, naturalist, philosopher, physician GERMANY

Arya Bàhram Picture Collection
Ernst Mach (1838 – 1916) physicist and philosopher
AUSTRIA

Ernst Mayr (1904 – 2005) evolutionary biologists, taxonomist, tropical explorer, ornithologist, and historian of science GERMANY

Arya Bàhram Picture Collection

Ernst Werner von Siemens (1816 – 1892) inventor and industrialist

Arya Bàhram Picture Collection

GERMANY

Arya Bàhram Picture Collection
Erwin Schrödingers
(1887 – 1961) physicist
AUSTRIA

Arya Bàhram Picture Collection

Euclid of Megara (435 – c. 365 BCE) philosopher

GREECE

Arya Bàhram Picture Collection
Eugene Paul Wigner (1902 – 1995) theoretical physicist and mathematician HUNGARY

Arya Bàhram Picture Collection
Eugenio Beltrami (1835 – 1899) mathematician
ITALY

Arya Bàhram Picture Collection
Evangelista Torricelli (1608 – 1647) physicist and mathematician ITALY

Arya Bàhram Picture Collection
Évariste Galois (1811 – 1832)   mathematician
FRANCE

Arya Bàhram Picture Collection — Fay Ajzenberg Selove (1926 – 2012) nuclear physicist GERMANY

Arya Bàhram Picture Collection
Felix Bloch (1905 – 1983) physicist
SWITZERLAND

Arya Bàhram Picture Collection
Félix Édouard Justin Émile Borel (1871 – 1956)
mathematician FRANCE

Arya Bàhram Picture Collection
Felix Hausdorff (1868 – 1942) mathematician
GERMANY

Arya Bàhram Picture Collection
Felix Hoffmann (1868 – 1946) chemist, credited for the first synthesized medically useful forms of heroin and aspirin GERMANY

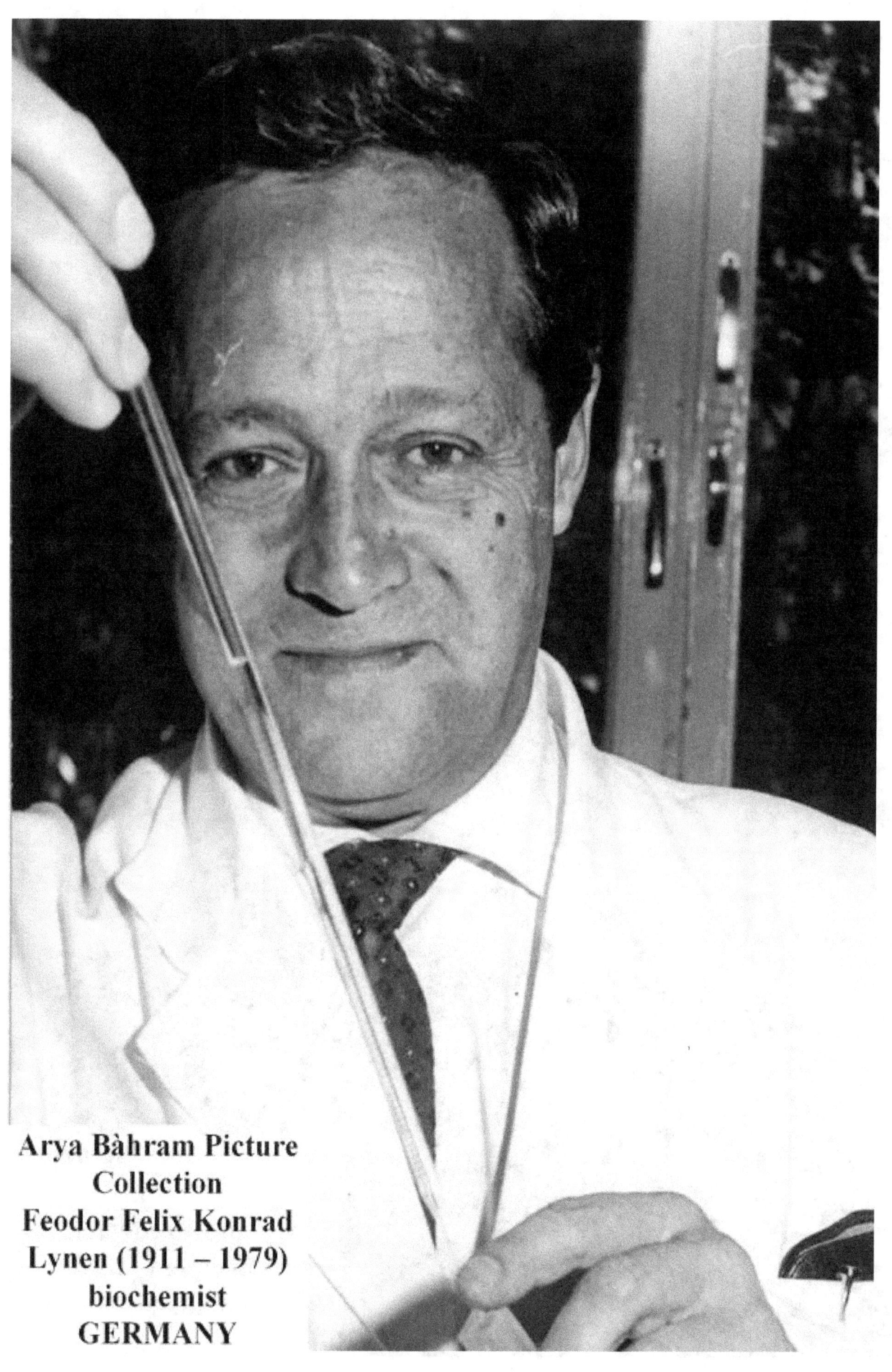

Arya Bàhram Picture Collection
Feodor Felix Konrad Lynen (1911 – 1979)
biochemist
GERMANY

Arya Bàhram Picture Collection
Ferdinand Frederick Henri Moissan (1852 – 1907) chemist
FRANCE

Arya Bàhram Picture Collection
Ferdinand Georg Frobenius (1849 – 1917) mathematician
GERMANY

Arya Bàhram Picture Collection
Ferdinand Gotthold Max Eisenstein (1823 – 1852)
mathematician   GERMANY

Arya Bàhram Picture Collection
Ferdinand Verbiest (1623 – 1688) mathematician and astronomer BELGIUM

Arya Bàhram Picture Collection
Francesco Bonaventura de Cavalieri (1598 – 1647)
mathematician ITALY

Arya Bàhram Picture Collection
Francesco Redi (1626 – 1697)   physician, naturalist
ITALY

Arya Bàhram Picture Collection
Francis Bacon (1561 – 1626) philosopher, scientist, jurist
ENGLAND

Arya Bàhram Picture Collection
Francis Crick (1916 – 2004) molecular biologist, biophysicist, and neuroscientist ENGLAND

Arya Bàhram Picture Collection
Francis Galton (1822 – 1911) psychologist, anthropologist, eugenicist, tropical explorer, geographer, inventor, meteorologist, proto-geneticist, psychometrician, and statistician   ENGLAND

Arya Bàhram Picture Collection
Francis William Aston (1877 – 1945) chemist and physicist
ENGLAND

Arya Bàhram Picture Collection
Franciscus Sylvius (1614 – 1672) chemical biologists, introduced the idea of chemical affinity to explain the human body's use of salts. He and his followers contributed greatly to the study of digestion and body fluids NETHERLANDS

Arya Bàhram Picture Collection — Franco Modigliani (1918 – 2003) — economist — ITALY — NOBEL PRIZE 1985

Arya Bahram Picture Collection
François Auguste Victor Grignard (1871 – 1935) chemist
France

Arya Bàhram Picture Collection
François Jacob (1920 – 2013) biologist
France

Arya Bahram Picture Collection
François Raoult (1830 – 1901) chemist who conducted research into the behavior of solutions, especially their physical properties FRANCE

Arya Bàhram Picture Collection
François Viète (1540 – 1603) mathematician
France

Arya Bàhram Picture Collection
Franz Boas (1858 – 1942) anthropologist
GERMANY

Arya Bàhram Picture Collection
Frederick Gowland Hopkins (1861 – 1947)   biochemist
ENGLAND

Arya Bàhram Picture Collection
Frederick Sanger (1918 – 2013) biochemist
ENGLAND

Arya Bàhram Picture Collection
Frederick Soddy (1877 – 1956)   radiochemist
ENGLAND

Arya Bàhram Picture Collection
Friedrich August Kekulé (1829 – 1896) organic chemist
GERMANY

Arya Bàhram Picture Collection
Friedrich Hayek (1899 – 1992) economist
AUSTRIA

Arya Bahram Picture Collection
Friedrich Karl Rudolf Bergius (1884 – 1949) chemist
GERMANY

Arya Bàhram Picture Collection
Friedrich Ludwig Gottlob Frege (1848 – 1925)
mathematician, logician and philosopher
GERMANY

Arya Bàhram Picture Collection
Friedrich Wöhler (1800 – 1882) chemist
GERMANY

Arya Bàhram Picture Collection
Frits Zernike (1888 – 1966) physicist
NETHERLANDS

Arya Bàhram Picture Collection
Fritz Albert Lipmann (1899 – 1986) biochemist
GERMANY

Arya Bàhram Picture Collection
Fritz Haber (1868 – 1934) chemist
GERMANY

Arya Bàhram Picture Collection
Fritz Pfleumer (1881 – 1945)   engineer who invented magnetic tape for recording sound   AUSTRIA

Arya Bàhram Picture Collection
Fritz Pregl (1869 – 1930) chemist and physician
SLOVENIA

Arya Bàhram Picture Collection
Galileo Galilei (1564 – 1642) physicist, mathematician, engineer, astronomer, philosopher
ITALY

Arya Bàhram Picture Collection
Gaspard Monge, Comte de Péluse (1746 – 1818)
mathematician, the inventor of descriptive geometry
FRANCE

Geoffrey Wilkinson (1921 – 1996) chemist ENGLAND   Arya Bàhram Picture Collection

Arya Bàhram Picture Collection
Georg Cantor (1845 – 1918) mathematician
GERMANY

Arya Bàhram Picture Collection
Georg Friedrich Bernhard Riemann (1826 – 1866)
mathematician   GERMANY

Arya Bàhram Picture Collection
Georg Ohm (1789 – 1854) physicist and mathematician
GERMANY

Sars discovered a new species of mysids and ostracods. His fileds of research included various branches of fishery.

Georg Sars (1837 – 1927) marine taxonomist.

Arya Bàhram Picture Collection

NORWAY

Arya Bàhram Picture Collection
Georg Wittig (1897 – 1987) chemist
GERMANY

Arya Bàhram Picture Collection
George Boole (1815 – 1864) mathematician, philosopher and logician ENGLAND

Arya Bàhram Picture Collection
George Charles de Hevesy (1885 – 1966) radiochemist
HUNGARY

Arya Bàhram Picture Collection — George Emil Palade (1912 – 2008) cell biologist ROMANIA

Arya Bàhram Picture Collection
George Gamow (1904 – 1968)   theoretical physicist and cosmologist RUSSIA

Arya Bàhram Picture Collection
George Hornidge Porter (1920 – 2002) chemist
ENGLAND

Arya Bàhram Picture Collection
George Paget Thomson (1892 – 1975) physicist
ENGLAND

Arya Bàhram Picture Collection

Georges Charpak
(1924 – 2010) physicist
POLAND

Arya Bàhram Picture Collection
Georges Louis Leclerc, Comte de Buffon (1707 – 1788)
naturalist, mathematician, cosmologist, and
encyclopedic author   FRANCE

Arya Bàhram Picture Collection Gérard Debreu (1921–2004) economist and mathematician FRANCE

Arya Bàhram Picture Collection
Gerhard Johannes Paul Domagk (1895 – 1964) pathologist and bacteriologist   GERMANY

Arya Bàhram Picture Collection
Germain Hess (1802 – 1850) chemist and doctor who formulated Hess's law, an early principle of thermochemistry SWITZERLAND

Arya Bàhram Picture Collection
Gerty Theresa Cori (1896 – 1957) biochemist
CZECH REPUBLIC

Giordano Bruno (1548 – 1600) philosopher, mathematician, poet, and astrologer. He is celebrated for his cosmological theories
Arya Bàhram Picture Collection
ITALY

Arya Bàhram Picture Collection
Girolamo Cardano (1501 – 1576) mathematician, physician, astrologer, philosopher ITALY

Arya Bàhram Picture Collection
Giulio Natta (1903 – 1979) chemist
ITALY

Arya Bàhram Picture Collection
Giuseppe Peano (1858 – 1932)   mathematician
ITALY

Arya Bàhram Picture Collection
Godfrey Harold Hardy
(1877 – 1947)
mathematician
ENGLAND

Arya Bàhram Picture Collection — Gottfried Wilhelm Leibniz (1646 – 1716) mathematician and philosopher GERMANY

Arya Bahram Picture Collection
Gottlieb Daimler (1834 – 1900) engineer, industrial designer GERMANY

Arya Bàhram Picture Collection
Gregor Mendel (1822 – 1884) scientist (founder of the modern genetics) CZECH REPUBLIC

Arya Bàhram Picture Collection
Guglielmo Marconi (1874 – 1937) inventor and electrical engineer ITALY

Arya Bàhram Picture Collection
Gustav Kirchoff (1824 – 1887) physicist
GERMANY

Arya Bàhram Picture Collection
Gustav Ludwig Hertz (1887 – 1975) experimental physicist GERMANY

Arya Bàhram Picture Collection
Gustav Weißkopf (1874 – 1927) engineer (designed and built gliders, flying machines and engines)
GERMANY

Arya Bàhram Picture Collection
György Békésy (1899 – 1972)   biophysicist
HUNGARY

Arya Bàhram Picture Collection
Hannes Olof Gösta Alfvén (1908 – 1995) electrical engineer, plasma physicist SWEDEN

Arya Bàhram Picture Collection
Hans Adolf Krebs (1900 – 1981) physician and biochemist
GERMANY

Arya Bàhram Picture Collection
Hans Bethe (1906 – 2005) nuclear physicist
GERMANY

Arya Bàhram Picture Collection
Hans Christian Oersted (1777 – 1851) physicist and chemist DENMARK

Arya Bàhram Picture Collection
Hans Fischer (1881 – 1945) organic chemist
GERMANY

Arya Bàhram Picture Collection
Hans Karl August Simon von Euler Chelpin (1873 – 1964)
biochemist   GERMANY

Arya Bàhram Picture Collection
Hans Lipperhey (1570 – 1619)   inventor (telescope)
GERMANY

Arya Bàhram Picture Collection
Hans Selye (1907 – 1982) endocrinologist

AUSTRIA

Arya Bàhram Picture Collection
Hans Spemann (1869 – 1941) embryologist
GERMANY

Arya Bàhram Picture Collection
Hans von Ohain (1911 – 1998) engineer designer of the first operational jet engine GERMANY

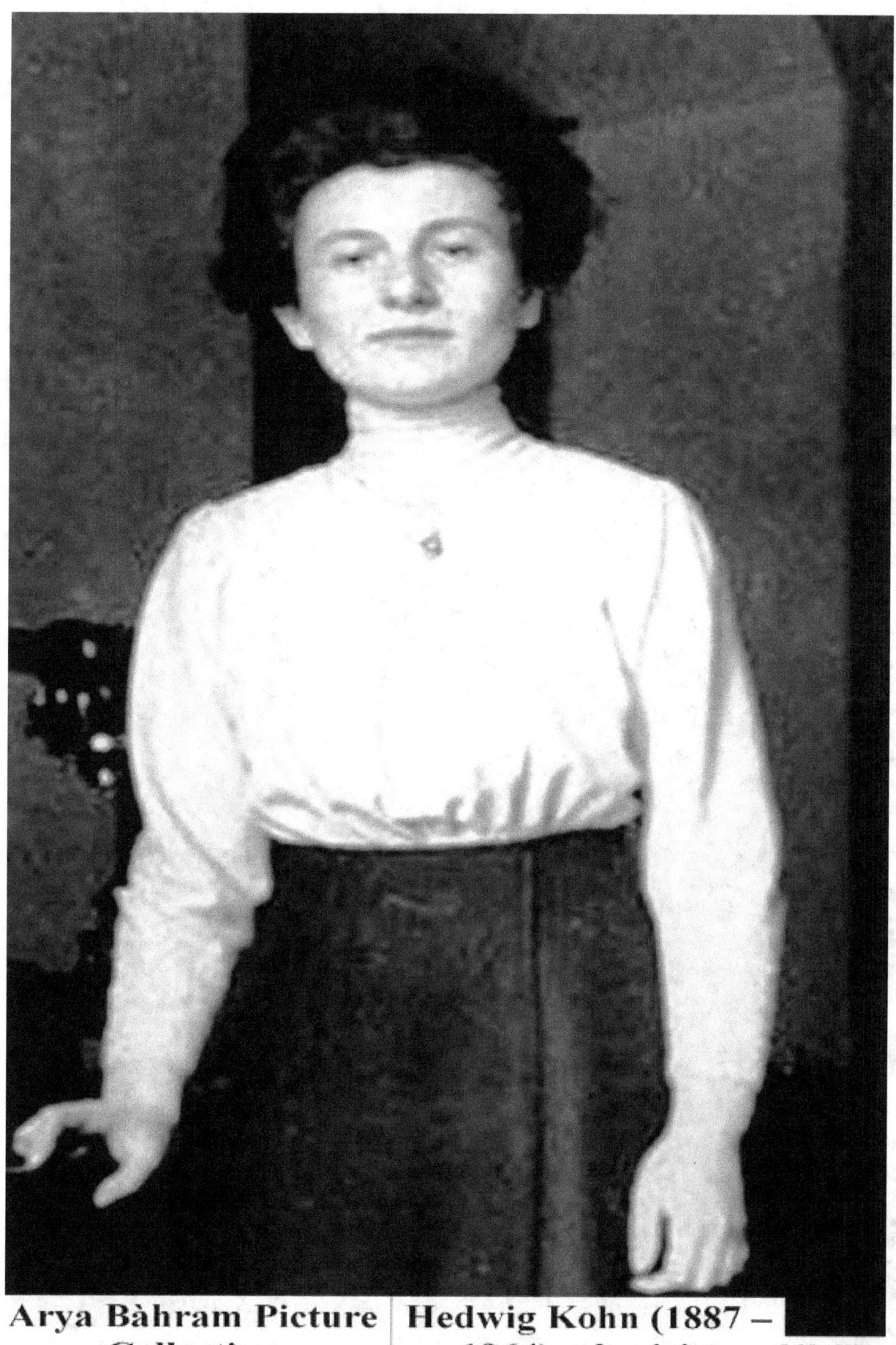

Arya Bàhram Picture Collection | **Hedwig Kohn (1887 – 1964) physicist** | POLAND

Arya Bahram Picture Collection
Heike Kamerlingh Onnes (1853 – 1926) physicist
NETHERLANDS

Arya Bàhram Picture Collection
Heinrich Göbel (1818 – 1893)   precision mechanic and inventor (Incandescent light bulb) GERMANY

Arya Bahram Picture Collection
Heinrich Hertz (1857 – 1894) physicist
GERMANY

Arya Bàhram Picture Collection
Heinrich Otto Wieland (1877 – 1957) chemist
GERMANY

Arya Bàhram Picture Collection

Heinrich Rohrer (1933 – 2013) physicist
SWITZERLAND

Arya Bàhram Picture Collection
Hendrik Antoon Lorentz (1853 – 1928) physicist
NETHERLANDS

Arya Bàhram Picture Collection
Henri Becquerel (1852 – 1908) physicist
FRANCE

Arya Bàhram Picture Collection

Henri Léon Lebesgue
(1875 – 1941)
mathematician

FRANCE

Arya Bàhram Picture Collection
Henrik Dam (1895 – 1976) biochemist and physiologist
DENMARK

Arya Bàhram Picture Collection
Henry Bessemer (1813 – 1898) Inventor (process for the manufacture of steel)   ENGLAND

Arya Bàhram Picture Collection
Henry Cole (1808 – 1882) inventor, Cole is credited with devising the concept of sending greetings cards at Christmas time, introducing the world's first commercial Christmas card in 1843

ENGLAND

Arya Bàhram Picture Collection
Henry Hallett Dale (1875 – 1968) pharmacologist and physiologist ENGLAND

Arya Bàhram Picture Collection
Henry John Stephen Smith (1826 – 1883) mathematician
IRELAND

Henry Louis Le Chatelier (1850 – 1936) chemist. He is most famous for devising Le Châtelier's principle   FRANCE

Arya Bàhram Picture Collection

Arya Bàhram Picture Collection
Henry Moseley (1887 – 1915)   physicist
ENGLAND

Arya Bàhram Picture Collection
Hermann Günter Grassmann (1809 – 1877)
mathematician, physicist, neohumanist, general scholar, and publisher GERMANY

Arya Bàhram Picture Collection
Hermann Klaus Hugo Weyl (1885 – 1955)
mathematician, theoretical physicist and philosopher
GERMANY

Arya Bàhram Picture Collection

mathematician Hermann Minkowski (1864 – 1909) GERMANY

Arya Bàhram Picture Collection
Hermann Oberth (1894 – 1989) physicist and engineer
GERMANY

Arya Bàhram Picture Collection
Hermann Rorschach (1884 – 1922) psychiatrist and psychoanalyst, SWITZERLAND

Arya Bàhram Picture Collection
Hermann Staudinger (1881 – 1965) chemist
GERMANY

Arya Bàhram Picture Collection
Hermann von Helmholtz (1821 – 1894) physician and physicist   GERMANY

Arya Bàhram Picture Collection — Hipparchus of Nicaea (ca 190 – 127 BC) astronomer, geographer, and mathematician GREECE

Arya Bàhram Picture Collection
Hippocrates of Kos ;(c. 460 – c. 370 BC) GREECE , also known as Hippocrates II, was a Greek physician of the Age of Pericles (Classical Greece), and is considered one of the most outstanding figures in the history of medicine

Arya Bàhram Picture Collection
Hugo Junkers (1859 – 1935) engineer and aircraft designer GERMANY

Arya Bahram Picture Collection
Hugo von Mohl (1805 – 1872)   botanist
GERMANY

Arya Bàhram Picture Collection — Humphry Davy (1778 – 1829) chemist and inventor — ENGLAND

Hypatia of Alexandria (AD 350 – 370; killed in 415) Neoplatonist philosopher

Arya Bàhram Picture Collection

GREECE

Arya Bàhram Picture Collection
Ida Noddack (1896 – 1978) chemist and physicist   GERMANY

Arya Bàhram Picture Collection
Igor Sikorsky (1889 – 1972) aviation pioneer in both helicopters and fixed-wing aircraft RUSSIA

Arya Bahram Picture Collection
Igor Yevgenyevich Tamm (1895 – 1971) physicist
RUSSIA

Arya Bàhram Picture Collection
Ilse Essers (1898 – 1994) engineer
GERMANY

Arya Bàhram Picture Collection
Ilya Ilyich Mechnikov (1845 – 1916) biologist, zoologist and protozoologist RUSSIA

Arya Bahram Picture Collection
Ilya Mikhailovich Frank (1908 – 1990) physicist
RUSSIA

Arya Bàhram Picture Collection
Ilya Romanovich Prigogine (1917 – 2003) physical chemist
RUSSIA

Arya Bàhram Picture Collection
Irene Joliot Curie (1897 – 1956) chemist, scientist
FRANCE

Arya Bàhram Picture Collection
Isaac Barrow (1630 – 1677) Christian theologian, and mathematician who is generally given credit for his early role in the development of infinitesimal calculus
ENGLAND

Arya Bàhram Picture Collection
Isaac Newton (1642 – 1727) physicist and mathematician, natural philosopher ENGLAND

Arya Bàhram Picture Collection
Israel Moiseevich Gelfand (1913 – 2009)
mathematician   RUSSIA

Arya Bàhram Picture Collection
Ivan Pavlov (1849 – 1936) physiologist
RUSSIA

Arya Bàhram Picture Collection
Jacob Bernoulli (1654 – 1705)   mathematician
SWITZERLAND

Arya Bàhram Picture Collection
Jacobus Henricus van 't Hoff, Jr. (1852 – 1911) physical and organic chemist
NETHERLANDS

Arya Bàhram Picture Collection
Jacques Charles (1746 – 1823)   inventor, scientist, mathematician, and balloonist   FRANCE

Arya Bàhram Picture Collection
Jacques Lucien Monod
(1910 – 1976) biologist
FRANCE

Arya Bàhram Picture Collection
Jacques Salomon Hadamard (1865 – 1963)
mathematician FRANCE

Arya Bàhram Picture Collection
Jacques Yves Cousteau (1910 – 1997)

researcher and an ecologist who studied the lives of underwater animals and plants
FRANCE

Arya Bàhram Picture Collection
Jakob Johan Adolf Appellöf (1857 – 1921)   marine zoologist SWEDEN

Arya Bàhram Picture Collection
Jakob Steiner (1796 – 1863) mathematician
SWITZERLAND

Arya Bàhram Picture Collection
James Chadwick (1891 – 1974) physicist
ENGLAND

Arya Bàhram Picture Collection — James Clerk Maxwell (1831 – 1879) mathematical physicist — SCOTLAND

Arya Bàhram Picture Collection
James Edward Meade
(1907 – 1995)
economist ENGLAND

Arya Bàhram Picture Collection
James Franck (1882 – 1964) physicist
GERMANY

Arya Bàhram Picture Collection
James Gregory (1638 – 1675) mathematician and astronomer SCOTLAND

Arya Bàhram Picture Collection
James Hutton (1726 – 1797) geologist, physician, chemical manufacturer, naturalist, and experimental agriculturalist
SCOTLAND

Arya Bàhram Picture Collection
James Joseph Sylvester (1814 – 1897) mathematician
ENGLAND

Arya Bàhram Picture Collection
James Prescott Joule (1818 – 1889) physicist and brewer
ENGLAND

Arya Bàhram Picture Collection
James Watt (1736 – 1819) inventor and mechanical engineer SCOTLAND

Arya Bàhram Picture Collection
James Whyte Black (1924 – 2010) pharmacologist
SCOTLAND

Arya Bàhram Picture Collection
Jan Baptist van Helmont (1580 – 1644) and physician chemist, physiologist, BELGIUM

Arya Bàhram Picture Collection
Jan Tinbergen (1903 – 1994) economist
NETHERLANDS

Arya Bàhram Picture Collection
Janos Lajos (1903 – 1957) physicist, inventor mathematician, HUNGARY

Arya Bàhram Picture Collection
Jaroslav Heyrovský (1890 – 1967) chemist and inventor   CZECH REPUBLIC

Arya Bàhram Picture Collection
Jean Baptiste Gabriel Joachim Dausset (1916 – 2009)
immunologist FRANCE

Arya Bàhram Picture Collection
Jean Baptiste Joseph Fourier (1768 – 1830)
mathematician and physicist
FRANCE

Arya Bàhram Picture Collection
Jean Baptiste Lamarck (1744 – 1829) naturalist
France

Arya Bàhram Picture Collection
Jean Baptiste Perrin (1870 – 1942) physicist
France

Arya Bàhram Picture Collection
Jean Frédéric Joliot Curie (1900 – 1958) physicist
France

Arya Bàhram Picture Collection
Jean Gaston Darboux (1842 – 1917) mathematician
France

Arya Bàhram Picture Collection
Jean le Rond d'Alembert (1717 – 1783) mathematician, mechanician, physicist, philosopher and music theorist
FRANCE

Arya Bàhram Picture Collection
Jean Nollet (1700 – 1770) clergyman and physicist
FRANCE

Arya Bàhram Picture Collection
Jean Piaget (1896 – 1980) psychologist and philosopher SWITZERLAND

Arya Bahram Picture Collection
Jean Victor Poncelet (1788 – 1867) engineer and mathematician FRANCE

Arya Bàhram Picture Collection
Johann Bernoulli (1667 – 1748) mathematician
SWITZERLAND

Arya Bàhram Picture Collection
Johann Friedrich Wilhelm Adolf von Baeyer (1835 – 1917)
chemist GERMANY

Arya Bàhram Picture Collection
Johann Heinrich Lambert (1727 – 1777) mathematician, physicist, philosopher, astronomer
SWITZERLAND

Arya Bàhram Picture Collection
Johann Peter Gustav Lejeune Dirichlet (1805 – 1859)
mathematician GERMANY

Arya Bahram Picture Collection
Johann Wolfgang Dobereiner (1780 – 1849) chemist who is best known for work that foreshadowed the periodic law for the chemical elements   GERMANY

Arya Bàhram Picture Collection
Johannes Andreas Grib Fibiger (1867 – 1928) scientist, physician and professor of pathological anatomy
DENMARK

Arya Bahram Picture Collection
Johannes Diderik van der Waals (1837 – 1923) theoretical physicist and thermodynamicist
NETHERLANDS

Arya Bahram Picture Collection
Johannes Gutenberg (1398 – 1468) blacksmith, goldsmith, printer, and publisher GERMANY

Arya Bàhram Picture Collection
Johannes Hans Daniel Jensen (1907 – 1973) nuclear physicist GERMANY

Arya Bàhram Picture Collection
Johannes Kepler (1571 – 1630) mathematician, astronomer, and astrologer GERMANY

Arya Bàhram Picture Collection
Johannes Müller von Königsberg 'Regiomontanus' (1436 – 1476)   mathematician, astronomer, astrologer, translator, instrument maker   GERMANY

Arya Bahram Picture Collection
Johannes Nicolaus Brønsted (1879 – 1947) physical chemist DENMARK

Arya Bàhram Picture Collection
Johannes Stark (1874 – 1957) physicist
GERMANY

Arya Bàhram Picture Collection
John Bell (1763 – 1820)   anatomist and surgeon
SCOTLAND

Arya Bahram Picture Collection
John Charles Harsanyi (1920 – 2000) economist
HUNGARY

Arya Bàhram Picture Collection
John Cowdery Kendrew (1917 – 1997) biochemist and crystallographer ENGLAND

Arya Bahram Picture Collection
John Dalton (1766 – 1844) chemist, meteorologist and physicist ENGLAND

Arya Bàhram Picture Collection
John Douglas Cockcroft (1897 – 1967) physicist ENGLAND

Arya Bàhram Picture Collection
John James Rickard Macleod (1876 – 1935)   biochemist and physiologist   SCOTLAND

Arya Bàhram Picture Collection
John Locke (1632 – 1704) philosopher and physician
ENGLAND

Arya Bàhram Picture Collection
John Logie Baird (1888 – 1946) — engineer, innovator, inventor SCOTLAND

Arya Bàhram Picture Collection
John Murray (1841 – 1914) oceanographer and a marine biologist SCOTLAND

Arya Bàhram Picture Collection
John Napier (1550 – 1617) mathematician, physicist, and astronomer. SCOTLAND

Arya Bàhram Picture Collection
John Needham (1713 – 1781) biologist
ENGLAND

Arya Bàhram Picture Collection
John Ray (1627 – 1705) naturalist
ENGLAND

Arya Bàhram Picture Collection
John Richard Nicholas Stone (1913 – 1991)   economist
ENGLAND

Arya Bàhram Picture Collection
John Robert Vane (1927 – 2004) pharmacologist
ENGLAND

Arya Bàhram Picture Collection
John Stewart Bell (1928 – 1990) physicist, and the originator of Bell's theorem, a significant theorem in quantum physics regarding hidden variable theories
IRELAND

Arya Bâhram Picture Collection
John Venn (1834 – 1923) logician and philosopher noted for introducing the Venn diagram, used in the fields of set theory, probability, logic, statistics, and computer science
ENGLAND

Arya Bàhram Picture Collection
John Wallis (1616 – 1703) mathematician
ENGLAND

Arya Bàhram Picture Collection
John William Strutt, 3rd Baron Rayleigh (1842 – 1919)
physicist ENGLAND

Arya Bàhram Picture Collection
Jonas Ferdinand Gabriel Lippmann (1845 – 1921)
physicist and inventor   LUXEMBOURG

Arya Bàhram Picture Collection
Jöns Jacob Berzelius (1779 – 1848) chemist, one of the founders of modern chemistry   SWEDEN

Arya Bàhram Picture Collection
Joseph Banks (1743 – 1820) naturalist, botanist
ENGLAND

Arya Bàhram Picture Collection
Joseph Bertrand (1822 – 1900) mathematician who worked in the fields of number theory, differential geometry, probability theory, economics and thermodynamics
FRANCE

Arya Bàhram Picture Collection
Joseph John Thomson (1856 – 1940) physicist
ENGLAND

Arya Bàhram Picture Collection
Joseph Liouville (1809 – 1882)
mathematician   FRANCE

Arya Bàhram Picture Collection
Joseph Lister (1827 – 1912) surgeon and a pioneer of antiseptic surgery. ENGLAND

Arya Bàhram Picture Collection
Joseph Louis Gay Lussac (1778 – 1850) chemist and physicist. He is known mostly for two laws related to gases, and for his work on alcohol-water mixtures, which led to the degrees Gay-Lussac used to measure alcoholic beverages in many countries   FRANCE

Arya Bàhram Picture Collection
Joseph Louis Lagrange (1736 – 1813) mathematician, astronomer ITALY

Arya Bàhram Picture Collection
Joseph Priestley (1733 – 1804)   theologian, Dissenting clergyman, philosopher   ENGLAND

Arya Bàhram Picture Collection
Joseph von Fraunhofer (1787 – 1826)   optician
GERMANY

Arya Bàhram Picture Collection
Jules Henri Poincaré (1854 – 1912) mathematician, theoretical physicist, engineer, and a philosopher of science
FRANCE

Arya Bàhram Picture Collection
Jules Jean Baptiste Vincent Bordet (1870 – 1961)
immunologist and microbiologist
BELGIUM

Arya Bàhram Picture Collection
Julius Lothar Meyer (1830 – 1895) chemist, First person to draw the periodic table of chemical elements
GERMANY

Arya Bàhram Picture Collection
Julius Plücker (1801 – 1868) mathematician and physicist
GERMANY

Arya Bàhram Picture Collection
Julius Robert von Mayer (1814 – 1878) physician and physicist and one of the founders of thermodynamics
GERMANY

Arya Bàhràm Picture Collection
Julius Wagner Jauregg (1857 – 1940) physician
AUSTRIA

Arya Bàhram Picture Collection
Justus von Liebig
(1803 – 1873) chemist
GERMANY

Arya Bàhram Picture Collection
Karl Ernst von Baer (1792 – 1876) naturalist, biologist, geologist, meteorologist, geographer
ESTONIA

Arya Bàhram Picture Collection
Karl Ferdinand Braun (1850 – 1918) inventor, physicist
GERMANY

Arya Bàhram Picture Collection
Karl Friedrich von Drais (1785 – 1851) inventor, he invented the Laufmaschine (running machine), also later called the velocipede GERMANY

Arya Bàhram Picture Collection
Karl Gunnar Myrdal
(1898 – 1987)
economist, sociologist   SWEDEN

Arya Bahram Picture Collection
Karl Landsteiner (1868 – 1943) biologist and physician
AUSTRIA

Arya Bàhram Picture Collection
Karl Manne Georg Siegbahn (1886 – 1978)
physicist   SWEDEN

Arya Bàhram Picture Collection
Karl Ritter von Frisch
(1886 – 1982)
ethologist   AUSTRIA

Arya Bàhram Picture Collection
Karl Sune Detlof Bergström (1916 – 2004) biochemist
SWEDEN

Arya Bàhram Picture Collection
Karl Waldemar Ziegler (1898 – 1973) chemist
GERMANY

Arya Bàhram Picture Collection

GERMANY

Karl Wilhelm Theodor Weierstrass (1815 – 1897) mathematician

Arya Bàhram Picture Collection

Konrad Emil Bloch (1912 – 2000) biochemist

GERMANY

Arya Bàhram Picture Collection
Konrad Lorenz (1903 – 1989)   zoologist, ethologist, and ornithologist   AUSTRIA

Arya Bàhram Picture Collection

Konrad Zuse (1910 – 1995) engineer, inventor and computer pioneer GERMANY

Arya Bàhram Picture Collection
Kristian Birkeland (1867 – 1917) scientist
NORWAY

Arya Bàhram Picture Collection
Kurt Alder (1902 – 1958) chemist
GERMANY

Kurt Gödel (1906 – 1978) logician, mathematician, and philosopher
AUSTRIA

Arya Bàhram Picture Collection

Arya Bàhram Picture Collection
Lars Onsager (1903 – 1976) physical chemist and theoretical physicist   NORWAY

Arya Bàhram Picture Collection
Lars Valerian Ahlfors (1907 – 1996) mathematician, remembered for his work in the field of Riemann surfaces and his text on complex analysis   FINLAND

Arya Bàhram Picture Collection
Lars Valter Hörmander (1931 – 2012) mathematician who has been called "the foremost contributor to the modern theory of linear partial differential equations"
SWEDEN

Arya Bàhram Picture Collection
Laszlo Jozsef Biro (1899 – 1985) inventor, the ballpoint pen, still commonly called biro after him.
HUNGARY

Arya Bàhram Picture Collection
Laura Bassi (1711 – 1778) scientist, physicist
ITALY

Arya Bàhram Picture Collection
Leo Baekeland (1863 – 1944) chemist. He invented Velox
photographic paper in 1893 and Bakelite in 1907
BELGIUM

Arya Bàhram Picture Collection
Leo Szilard (1898 – 1964) physicist and inventor
HUNGARY

Arya Bàhram Picture Collection
Leon Foucault (1819 – 1868) physicist
FRANCE

| Arya Bàhram Picture Collection | Leonard Fuchs (1501 – 1566) | physician and botanist GERMANY |

Arya Bàhram Picture Collection
Leonardo da Vinci (1452 – 1519) architect,
mathematician, engineer, inventor, anatomist, geologist,
cartographer, botanist, writer, painter, sculptor, musician
ITALY

Arya Bàhram Picture Collection
Leonhard Euler (1707 – 1783)   mathematician and physicist   SWITZERLAND

Arya Bàhram Picture Collection
Leonid Hurwicz (1917 – 2008)   economist and mathematician, He originated incentive compatibility and mechanism design   RUSSIA

Arya Bahram Picture Collection
Leopold Kronecker (1823 – 1891) mathematician
GERMANY

Arya Bàhram Picture Collection
Leopold Ružička (1887 – 1976) scientist, chemist
CROATIA

Arya Bàhram Picture Collection
Lev Davidovich Landau (1908 – 1968) physicist
RUSSIA

Arya Bàhram Picture Collection
Lise Meitner (1878 – 1968) physicist
AUSTRIA

Arya Bàhram Picture Collection
Louis Agassiz (1807 – 1873) biologist and geologist
SWITZERLAND

Arya Bàhram Picture Collection
Louis Braille (1809 – 1852) educator and inventor of a system of reading and writing for use by the blind or visually impaired. His system remains known worldwide simply as Braille   FRANCE

Arya Bàhram Picture Collection
Louis de Broglie (1892 – 1987)   physicist
FRANCE

Arya Bàhram Picture Collection
Louis Pasteur (1822 – 1895) chemist and microbiologist
FRANCE

Arya Bàhram Picture Collection
Lucretius (99 BC – c. 55 BC) philosopher
ITALY

Arya Bàhram Picture Collection
Ludwig Boltzmann (1844 – 1906) physicist and philosopher AUSTRIA

Arya Bàhram Picture Collection
Luigi Galvani (1737 – 1798) physicist and philosopher
ITALY

Arya Bàhram Picture Collection
Manfred von Ardenne (1907 – 1997) physicist and inventor
GERMANY

Arya Bàhram Picture Collection
Marcello Malpighi (1628 – 1694) physician and biologist
ITALY

Arya Bàhram Picture Collection Marcus Terentius Varro (116 BC – 27 BC) ITALY, he was an ancient Roman scholar and writer.

Arya Bàhram Picture Collection
Maria Gaetana Agnesi (1718 – 1799) mathematician and philosopher ITALY

Arya Bàhram Picture Collection
Maria Goeppert Mayer (1906 – 1972) theoretical physicist
GERMANY

Arya Bàhram Picture Collection
Maria Telkes (1900 – 1995) scientist and inventor (solar energy technologies) HUNGARY

Arya Bàhram Picture Collection
Marie Curie (1867 – 1934) physicist and chemist
POLAND

Arya Bàhram Picture Collection
Marie Ennemond Camille Jordan (1838 – 1922)
mathematician FRANCE

Arya Bàhram Picture Collection
Marius Sophus Lie (1842 – 1899) mathematician
NORWAY

Arya Bàhram Picture Collection
Mary Anning (1799 – 1847) fossil collector, dealer, and palaeontologist ENGLAND

Arya Bàhram Picture Collection
Mary Cartwright (1900 – 1998) mathematician, she was the first to analyze a dynamical system with chaos
ENGLAND

Arya Bàhram Picture Collection
Mary Somerville (1780 – 1872) science writer and polymath, at a time when women's participation in science was discouraged. She studied mathematics and astronomy
SCOTLAND

Arya Bàhram Picture Collection
Maurice Félix Charles Allais (1911 – 2010)   economist
FRANCE

Arya Bàhram Picture Collection
Maurice Hugh Frederick Wilkins (1916 – 2004) physicist, molecular biologist   ENGLAND

Arya Bàhram Picture Collection
Max Born (1882 – 1970) physicist and mathematician
GERMANY

Arya Bàhram Picture Collection
Max Ferdinand Perutz
(1914 – 2002)
molecular biologist
AUSTRIA

Arya Bàhram Picture Collection
Max Planck (1858 – 1947) theoretical physicist
GERMANY

Arya Bàhram Picture Collection
Max von Laue (1879 – 1960) physicist
GERMANY

Arya Bàhram Picture Collection
Melitta Bentz (1873 – 1950) inventor (the coffee filter), entrepreneur   GERMANY

Arya Bàhram Picture Collection
Michael Faraday (1791 – 1867) scientist
(electromagnetism, electrochemistry)
ENGLAND

Arya Bàhram Picture Collection
Mihailo Petrovic Alas (1868 – 1943) mathematician and inventor SERBIA

Arya Bàhram Picture Collection
Nevill Francis Mott (1905 – 1996) physicist
ENGLAND

Arya Bàhram Picture Collection
Nicephore Niepce (1765 – 1833) inventor, now usually credited as the inventor of photography and a pioneer in that field   FRANCE

Arya Bàhram Picture Collection
Nicholai Maklai (1846 – 1888) anthropologist, ethnologist and a marine biologist who graduated from St. Petersburg University RUSSIA

Arya Bàhram Picture Collection
Nicholas Culpeper (1616 – 1654) botanist, herbalist, physician, and astrologer   ENGLAND

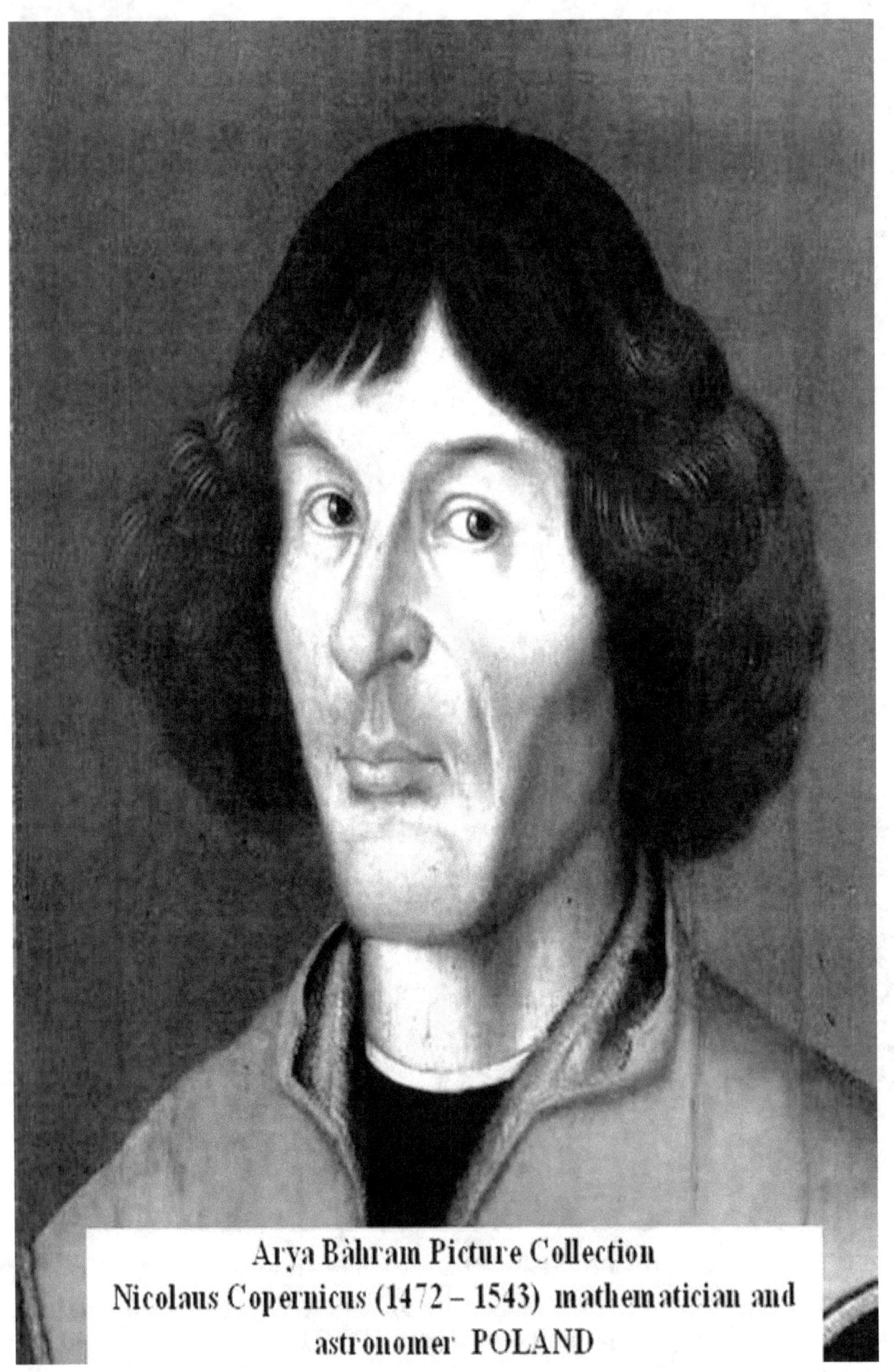

Arya Bàhram Picture Collection
Nicolaus Copernicus (1472 – 1543) mathematician and astronomer POLAND

Arya Bàhram Picture Collection
Nicole Oresme (ca 1322 – 1382) philosopher (He wrote works on economics, mathematics, physics, astrology and astronomy, philosophy, and theology)   FRANCE

Arya Bàhram Picture Collection
Niels Bohr (1885 – 1962) physicist
DENMARK

Arya Bàhram Picture Collection
Niels Henrik Abel (1802 – 1829) mathematician
NORWAY

Arya Bàhram Picture Collection
Niels Kaj Jerne (1911 – 1994)   immunologist
DENMARK

Arya Bàhram Picture Collection
Niels Ryberg Finsen (1860 – 1904) physician and scientist
FAROE ISLANDS

Arya Bàhram Picture Collection
Nikola Tesla (1856 – 1943) inventor, electrical engineer, mechanical engineer, SERBIA

Arya Bàhram Picture Collection
Nikolaas Tinbergen (1907 – 1988) biologist and ornithologist
NETHERLANDS

Arya Bàhram Picture Collection
Nikolai Ivanovich Lobachevsky (1793 – 1856)
mathematician and geometer   RUSSIA

Arya Bàhram Picture Collection
Nikolay Nikolayevich Semyonov (1896 – 1986) physicist and chemist   RUSSIA

Arya Bàhram Picture Collection
Nils Gustaf Dalén (1869 – 1937) scientist, inventor
SWEDEN

Arya Bàhram Picture Collection
Odd Hassel (1897 – 1981) physical chemist
NORWAY

Arya Bàhram Picture Collection
Oliver Heaviside (1850 – 1925) electrical engineer, mathematician, physicist   ENGLAND

Arya Bàhram Picture Collection
Otto Brunfels (1488 – 1534) theologian and botanist, was the first to produce a major work on plants
GERMANY

Arya Bàhram Picture Collection
Otto Fritz Meyerhof (1884 – 1951) physician and biochemist GERMANY

Arya Bàhram Picture Collection
Otto Hahn (1879 – 1968) chemist and pioneer in the fields of radioactivity and Radiochemistry
GERMANY

Arya Bàhram Picture Collection
Otto Haxel (1909 – 1998) nuclear physicist
GERMANY

Arya Bàhram Picture Collection
Otto Heinrich Warburg (1883 – 1970) physiologist, medical doctor   GERMANY

Arya Bàhram Picture Collection
Otto Lilienthal (1848 – 1896) pioneer of aviation
GERMANY

Arya Bàhram Picture Collection
Otto Loewi (1873 – 1961) pharmacologist
GERMANY

Arya Bàhram Picture Collection
Otto Paul Hermann Diels (1876 – 1954) chemist
GERMANY

Arya Bàhram Picture Collection
Otto Stern (1888 – 1969) physicist
GERMANY

Arya Bàhram Picture Collection
Otto von Guericke (1602 – 1686) scientist, inventor the air pump, and did the first experiments with vacuums. GERMANY

Arya Bàhram Picture Collection
Otto Wallach (1847 – 1931) chemist
GERMANY

Arya Bàhram Picture Collection
Owen Willans Richardson (1879 – 1959) physicist
ENGLAND

Arya Bàhram Picture Collection
Pafnuty Lvovich Chebyshev (1821 – 1894)
mathematician   RUSSIA

Arya Bàhram Picture Collection
Paracelsus (1493 – 1541) physician, botanist, alchemist,
astrologer, occultist   SWITZERLAND

Arya Bàhram Picture Collection
Patrick Maynard Stuart Blackett (1897 – 1974) experimental physicist
ENGLAND

Arya Bàhram Picture Collection
Paul Dirac (1902 – 1984) theoretical physicist
ENGLAND

Arya Bàhram Picture Collection
Paul Ehrlich (1854 – 1915) physician and scientist who worked in the fields of hematology, immunology, and chemotherapy   GERMANY

Arya Bàhram Picture Collection

Paul Erdös (1913 – 1996) mathematician. He was the most prolific mathematician of the 20th century
HUNGARY

Arya Bahram Picture Collection
Paul Karrer (1889 – 1971) organic chemist
SWITZERLAND

Arya Bàhram Picture Collection
Paul Sabatier (1854 – 1941) chemist
FRANCE

Arya Bàhram Picture Collection
Peter Brian Medawar (1915 – 1987) biologist
ENGLAND

Arya Bàhram Picture Collection
Peter Debye (1884 – 1966) physicist and physical chemist, NETHERLANDS

Arya Bàhram Picture Collection
Peter Henlein (1485 – 1542) locksmith and clockmaker of Nuremberg, Germany, is often considered the inventor of the watch   GERMANY

Arya Bàhram Picture Collection
Philipp Reis (1834 – 1874) scientist and inventor
GERMANY

Arya Bàhram Picture Collection
Pierre Curie (1859 – 1906) physicist, a pioneer in crystallography, magnetism, piezoelectricity and radioactivity   FRANCE

Arya Bàhram Picture Collection
Pierre de Fermat (1601 – 1665) mathematician
(infinitesimal calculus, including his technique of adequality) FRANCE

Arya Bàhram Picture Collection
Pierre Gilles de Gennes (1932 – 2007)
physicist
FRANCE

Arya Bàhram Picture Collection
Pierre Simon Laplace (1749 – 1827) mathematician and astronomer   FRANCE

Arya Bàhram Picture Collection
Pieter Zeeman (1865 – 1943) physicist
NETHERLANDS

Arya Bàhram Picture Collection
Plato (424/423 – 348/347 BCE) GREECE, He was a philosopher in Classical Greece and the founder of the Academy in Athens, the first institution of higher learning in the Western world

Arya Bàhram Picture Collection
Prokop Divis (1698 – 1765) theologian and natural scientist, who invented the first grounded lightning rod
CZECH REPUBLIC

Arya Bàhram Picture Collection
Pythagoras of Samos (c. 570 BC – c. 495 BC) philosopher, mathematician, and founder of the religious movement called Pythagoreanism
GREECE

Arya Bàhram Picture Collection
Ragnar Arthur Granit (1900 – 1991) scientist (Physiology, Medicine) FINLAND

Arya Bàhram Picture Collection
Renato Dulbecco (1914 – 2012) virologist
ITALY

Arya Bàhram Picture Collection
René Déscartes (1596 – 1650)   philosopher,
mathematician and writer   FRANCE

Arya Bàhram Picture Collection
Rene Laennec (1781 – 1826) physician. He invented the stethoscope in 1816, while working at the Hôpital Necker and pioneered its use in diagnosing various chest conditions FRANCE

Arya Bàhram Picture Collection
René Maurice Fréchet
(1878 – 1973)
mathematician
FRANCE

Arya Bàhram Picture Collection
Richard Adolf Zsigmondy (1865 – 1929) chemist
AUSTRIA

Arya Bàhram Picture Collection
Richard Kuhn (1900 – 1967) biochemist
AUSTRIA

Arya Bàhram Picture Collection
Richard Laurence Millington Synge (1914 - 1994)
biochemist   ENGLAND

Arya Bàhram Picture Collection
Richard Martin Willstätter (1872 – 1942) organic chemist
GERMANY

Arya Bàhram Picture Collection
Rita Levi Montalcini
(1909 – 2012)
neurologist
ITALY

Arya Bahram Picture Collection
Róbert Bárány (1876 – 1936) otologist
AUSTRIA

Arya Bàhram Picture Collection

Robert Bosch (1861 – 1942) engineer and inventor GERMANY

Arya Bàhram Picture Collection
Robert Boyle (1627 – 1691) natural philosopher, chemist, physicist, inventor   IRELAND

Arya Bàhram Picture Collection
Robert Brown (1773 – 1858) botanist and palaeobotanist
SCOTLAND

Arya Bàhram Picture Collection
Robert Bunsen (1811 – 1899)   chemist. He investigated
emission spectra of heated elements GERMANY

Arya Bàhram Picture Collection
Robert Hooke (1635 – 1703) natural philosopher, architect and polymath   ENGLAND

Arya Bàhram Picture Collection
Robert Koch (1843 – 1910) physician and pioneering microbiologist. The founder of modern bacteriology
GERMANY

Arya Bàhram Picture Collection
Robert Robinson (1886 – 1975) organic chemist
ENGLAND

Arya Bàhram Picture Collection
Rodney Robert Porter
(1917 – 1985)
biochemist
ENGLAND

Arya Bàhram Picture Collection
Roger Bacon (1214 – 1294) philosopher
ENGLAND

Arya Bàhram Picture Collection
Ronald George Wreyford Norrish (1897 – 1978) chemist
ENGLAND

**Arya Bàhram Picture Collection
Ronald Ross (1857 – 1932) medical doctor
ENGLAND**

**Arya Bàhram Picture Collection
Rosalind Franklin (1920 – 1958) chemist and X-ray
crystallographer ENGLAND**

Arya Bàhram Picture Collection
Rudolf Christian Karl Diesel (1858 – 1913)   inventor and mechanical engineer GERMANY

Arya Bàhram Picture Collection — Rudolf Virchow (1821 – 1902) doctor, anthropologist, pathologist, prehistorian, biologist, writer, editor GERMANY

Arya Bàhram Picture Collection
Samuel Hahnemann (1755 – 1843) physician, best known for creating a system of alternative medicine called homeopathy
GERMANY

Arya Bàhram Picture Collection
Santiago Ramón y Cajal (1852 – 1934) pathologist, histologist, neuroscientist   SPAIN

Arya Bàhram Picture Collection
Santorio Santorio (1561 – 1636) physiologist, physician, and professor. He introduced the quantitative approach into medicine and, as his pupil, introduced the mechanistic principles of Galileo Galilei to medicine   ITALY

Arya Bàhram Picture Collection
Severo Ochoa de Albornoz (1905 – 1993) physician and biochemist SPAIN

Arya Bàhram Picture Collection
Sigmund Freud (1856 – 1939)   neurologist
AUSTRIA

Arya Bàhram Picture Collection
Siméon Denis Poisson (1781 – 1840) mathematician, geometer, and physicist FRANCE

Arya Bàhram Picture Collection
Simon Stevin (1549 – 1620) mathematician and military engineer   BELGIUM

Arya Bàhram Picture Collection
Sofia Vasilyevna Kovalevskaya (1850 – 1891)
mathematician   RUSSIA

Arya Bàhram Picture Collection
Stanislao Cannizzaro (1826 – 1910) chemist. He is remembered today largely for the Cannizzaro reaction and for his influential role in the atomic-weight deliberations of the Karlsruhe Congress in 1860

ITALY

Arya Bàhram Picture Collection
Svante Arrhenius (1859 – 1927) scientist, physicist, chemist   SWEDEN

Arya Bàhram Picture Collection
Tadeusz Reichstein (1897 – 1996) chemist
POLAND

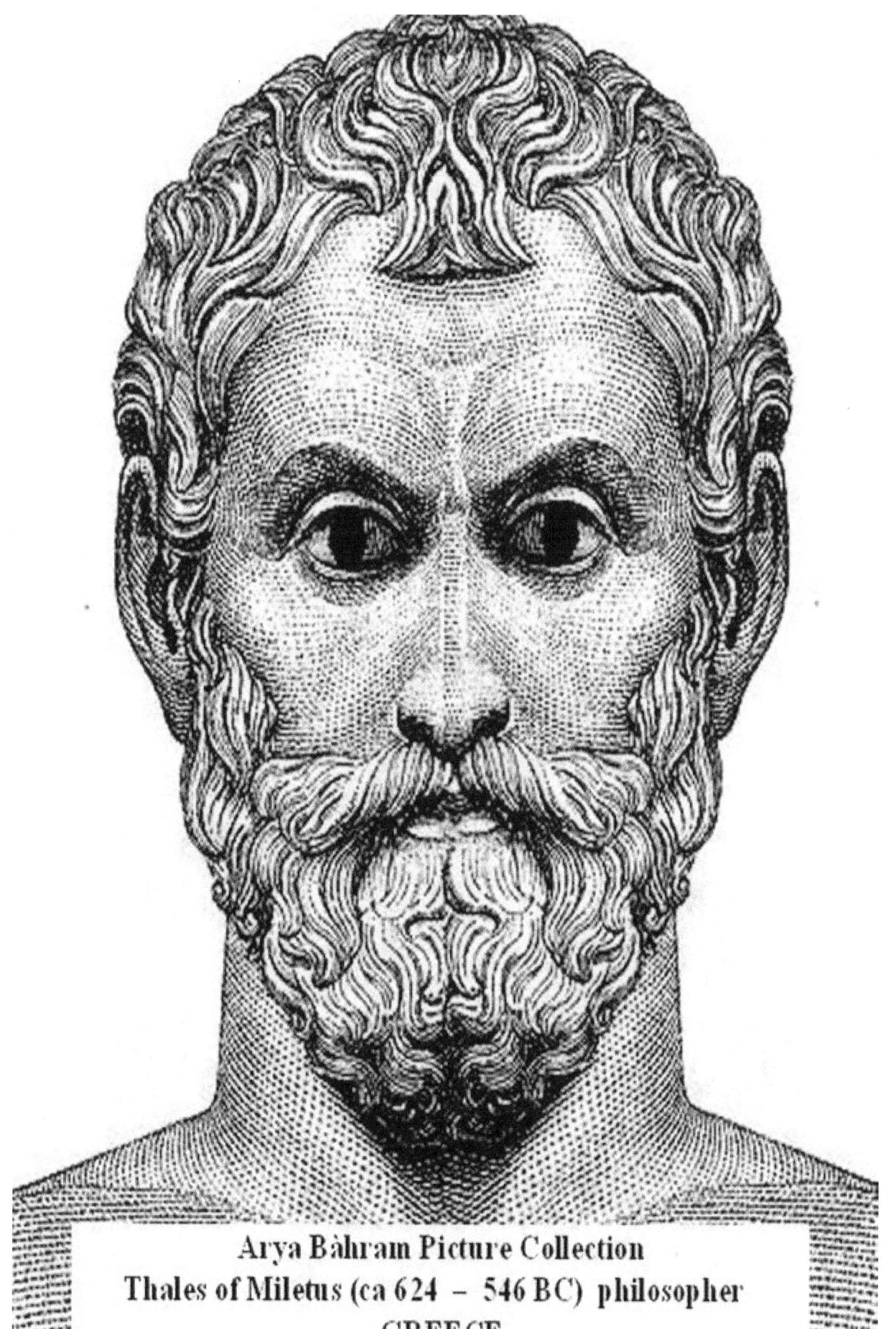

Arya Bàhram Picture Collection
Thales of Miletus (ca 624 – 546 BC) philosopher
GREECE

Arya Bàhram Picture Collection
Theodor Schwann (1810 – 1882) physiologist
GERMANY

Arya Bàhram Picture Collection
Theodor Svedberg (1884 – 1971) chemist
SWEDEN

Arya Bàhram Picture Collection
Thomas Burnet (1635? – 1715) theologian and writer on cosmogony   ENGLAND

Arya Bàhram Picture Collection
Thomas Graham (1805 – 1869) chemist. He developed a technique to separate crystalloids from colloids, which is called "dialysis"   SCOTLAND

Arya Bàhram Picture Collection
Thomas Martin Lowry (1874 – 1936) physical chemist who developed the Bronsted–Lowry acid–base theory simultaneously with and independently of Johannes Nicolaus Bronsted and was a founder-member and president (1928–1930) of the Faraday Society
ENGLAND

Arya Bàhram Picture Collection
Thomas Willis (1621 – 1675) doctor (who played an important part in the history of anatomy, neurology and psychiatry) ENGLAND

Arya Bàhram Picture Collection
Trofim Lysenko (1898 – 1976)   biologist and agronomist
RUSSIA

Arya Bàhram Picture Collection
Tullio Levi Civita (1873 – 1941) mathematician
ITALY

Arya Bàhram Picture Collection
Tycho Brahe (1546 – 1601) astronomer and alchemist and has been described more recently as "the first competent mind in modern astronomy to feel ardently the passion for exact empirical facts"   DENMARK

Arya Bàhram Picture Collection
Ulf Svante von Euler (1905 – 1983) physiologist and pharmacologist   SWEDEN

Arya Bàhram Picture Collection
Victor Francis Hess (1883 – 1964) physicist
AUSTRIA

Arya Bàhram Picture Collection
Vitaly Lazarevich Ginzburg (1916 – 2009)   theoretical physicist, astrophysicist   RUSSIA

Arya Bàhram Picture Collection
Vladimir Prelog (1906 – 1998)   organic chemist
CROATIA

Arya Bàhram Picture Collection
Vladimir Vernadsky (1863 – 1945) mineralogist and geochemist   RUSSIA

Arya Bàhram Picture Collection
Walter Hess (1881 – 1973) physiologist
SWITZERLAND

Arya Bàhram Picture Collection
Walter Schottky (1886 – 1976) physicist (who played a major early role in developing the theory of electron and ion emission phenomena) GERMANY

Arya Bàhram Picture Collection
Walther Hermann Nernst (1864 – 1941) physicist
GERMANY

Arya Bahram Picture Collection
Werner Heisenberg (1901 – 1976) theoretical physicist
and one of the key creators of quantum mechanics
GERMANY

Arya Bàhram Picture Collection
Werner Theodor Otto Forßmann (1904 – 1979) physician
GERMANY

Arya Bàhram Picture Collection
Wernher Von Braun (1912 – 1977) aerospace engineer
and space architect   GERMANY

Arya Bàhram Picture Collection
Wilhelm Bessel (1784 – 1846)   astronomer, mathematician (systematizer of the Bessel functions), He was the first astronomer to determine the distance from the sun to another star by the method of parallax
GERMANY

Arya Bàhram Picture Collection
Wilhelm Carl Werner Otto Fritz Franz Wien (1864 – 1928) physicist GERMANY

Arya Bàhram Picture Collection
Wilhelm Conrad Roentgen (1845 – 1923) physicist
GERMANY

Arya Bàhram Picture Collection
Wilhelm Ostwald (1853 – 1932) chemist
GERMANY

Arya Bàhram Picture Collection
Wilhelm Wundt (1832 – 1920) physician, physiologist, philosopher, professor GERMANY

Arya Bàhram Picture Collection
Willem Einthoven (1860 – 1927) doctor and physiologist
NETHERLANDS

Arya Bàhràm Picture Collection
William Bayliss (1860 – 1924)   physiologist
ENGLAND

Arya Bàhram Picture Collection
William Buckland (1784 – 1856) theologian, geologist and palaeontologist   ENGLAND

Arya Bàhram Picture Collection
William Friese Greene (1855 – 1921) portrait photographer and prolific inventor. He is principally known as a pioneer in the field of motion pictures and is credited by some as the "inventor" of cinematography.
ENGLAND

Arya Bàhram Picture Collection
William Harvey (1578 – 1657) physician
ENGLAND

Arya Bahram Picture Collection
William Henry Bragg (1862 – 1942) physicist, chemist, mathematician ENGLAND

Arya Bàhram Picture Collection
William Henry Perkin (1838 – 1907) chemist best known for his accidental discovery, at the age of 18, of the first aniline dye, mauveine. ENGLAND

Arya Bàhram Picture Collection
William Herschel (1738 – 1822) astronomer
GERMANY

Arya Bàhram Picture Collection
William Hopkins (1793 – 1866) mathematician and geologist   ENGLAND

Arya Bahram Picture Collection
William John Swainson (1789 – 1855) ornithologist, malacologist, conchologist, entomologist and artist
ENGLAND

Arya Bàhram Picture Collection
William Lawrence Bragg (1890 – 1971) physicist and X-ray crystallographer   ENGLAND

Arya Bàhram Picture Collection
William Ramsay (1852 – 1916) chemist
SCOTLAND

Arya Bàhram Picture Collection
William Rowan Hamilton (1805 – 1865) physicist, astronomer, and mathematician, who made important contributions to classical mechanics, optics, and algebra
IRELAND

Arya Bàhram Picture Collection
William Smith (1769 – 1839) geologist
ENGLAND

Arya Bàhram Picture Collection
William Thompson (1805 – 1852) naturalist celebrated for his founding studies of the natural history of Ireland, especially in ornithology and marine biology
IRELAND

Arya Bàhram Picture Collection
William Thomson (1824 – 1907) mathematical physicist and engineer IRELAND

Arya Bàhram Picture Collection
Wolfgang Ernst Pauli (1900 – 1958) theoretical physicist
and one of the pioneers of quantum physics
AUSTRIA

Arya Bàhram Picture Collection
Zacharias Janssen (1585 – 1632) spectacle-maker from Middelburg associated with the invention of the first optical telescope. Janssen is sometimes also credited for inventing the first truly compound microscope
NETHERLAND

www.ingramcontent.com/pod-product-compliance
Lightning Source LLC
Chambersburg PA
CBHW081137180526
45170CB00006B/1842